Social Philosophy after Adorno

This book examines what is living and what is dead in the social philosophy of Theodor W. Adorno, the most important philosopher and social critic in Germany after World War II. When he died in 1969, Adorno's successors abandoned his critical-utopian passions. Habermas, in particular, rejected or ignored Adorno's central insights on the negative effects of capitalism and new technologies upon nature and human life. In this book, Lambert Zuidervaart reclaims Adorno's insights from Habermasian neglect, while taking up legitimate Habermasian criticisms. He also addresses the prospects for radical and democratic transformations of an increasingly globalized world. The book proposes a provocative social philosophy "after Adorno."

Lambert Zuidervaart is Professor of Philosophy at the Institute for Christian Studies and an Associate Member of the Graduate Faculty in the Department of Philosophy at the University of Toronto. A specialist in hermeneutics, social theory, and German philosophy, he is the editor and author of several books, most recently *Artistic Truth: Aesthetics, Discourse, and Imaginative Disclosure*, which was selected by *Choice* as an Outstanding Academic Title for 2005 and received the Symposium Book Award from the Canadian Society for Continental Philosophy in 2006.

For Joyce, Esther, and Sophie
Sisters across three generations

Social Philosophy after Adorno

LAMBERT ZUIDERVAART
Institute for Christian Studies
University of Toronto

CAMBRIDGE
UNIVERSITY PRESS

32 Avenue of the Americas, New York NY 10013-2473, USA

Cambridge University Press is part of the University of Cambridge.

It furthers the University's mission by disseminating knowledge in the pursuit of education, learning and research at the highest international levels of excellence.

www.cambridge.org
Information on this title: www.cambridge.org/9780521690386

© Lambert Zuidervaart 2007

This publication is in copyright. Subject to statutory exception and to the provisions of relevant collective licensing agreements, no reproduction of any part may take place without the written permission of Cambridge University Press.

First published 2007

A catalogue record for this publication is available from the British Library

Library of Congress Cataloguing in Publication data

Zuidervaart, Lambert.
Social philosophy after Adorno / Lambert Zuidervaart.
    p. cm.
Includes bibliographical references and index.
ISBN 978-0-521-87027-6 (hardback) – ISBN 978-0-521-69038-6 (pbk.)
1. Adorno, Theodor W., 1903–1969. 2. Social sciences – Philosophy. I. Title.
B3199.A34Z84 2007
193–dc22   2007012541

ISBN 978-0-521-87027-6 Hardback
ISBN 978-0-521-69038-6 Paperback

Cambridge University Press has no responsibility for the persistence or accuracy of URLs for external or third-party internet websites referred to in this publication, and does not guarantee that any content on such websites is, or will remain, accurate or appropriate.

Contents

Preface		*page* vii
Abbreviations		xi
	Introduction: Thinking Otherwise	1
	1.1 *Wozu noch Philosophie?*	4
	1.2 *Going after Adorno*	7
	1.3 *Critical Retrieval*	10
1	Transgression or Transformation	16
	1.1 *Menke's Derridean Reconstruction*	18
	1.2 *Liberation and Deconstruction*	23
	1.3 *Aesthetic and Artistic Autonomy*	38
2	Metaphysics after Auschwitz	48
	2.1 *Wellmer's Postmetaphysical Critique*	49
	2.2 *Suffering, Hope, and Societal Evil*	58
	2.3 *Displaced Object*	66
3	Heidegger and Adorno in Reverse	77
	3.1 *Existential Authenticity*	78
	3.2 *Emphatic Experience*	95
	3.3 *Public Authentication*	101
4	Globalizing Dialectic of Enlightenment	107
	4.1 *Habermas's Paradigmatic Critique*	108
	4.2 *Remembrance of Nature*	112
	4.3 *Beyond Globalization*	125

5	Autonomy Reconfigured	132
	5.1 Feminist Cultural Politics	133
	5.2 The Culture Industry	137
	5.3 Culture, Politics, and Economy	145
6	Ethical Turns	155
	6.1 Adorno's Politics	157
	6.2 Social Ethics and Global Politics	162
	6.3 Resistance and Transformation	175

Appendix: Adorno's Social Philosophy — 183
Bibliography — 203
Index — 215

Preface

Matthew Klaassen and I were driving back to Toronto from the 2004 Critical Theory Roundtable in Montreal when I asked whether I should turn my recent work on Adorno into a book. Matt had just presented an excellent paper on Habermas's critique of Adorno, the topic of the master's thesis he would complete in 2005. He had attended my graduate seminars on Adorno's *Negative Dialectics* and on Habermas's *Theory of Communicative Action*. He had also been the research assistant for an encyclopedia entry on Adorno as well as for my book on *Artistic Truth*. No one else knew so well the themes of my recent research. So when Matt said yes, that was the signal I needed to begin a book on Adorno's social philosophy.

Substantial components had already been drafted. The earliest materials stem from an invited lecture for the Eslick Symposium at St. Louis University in 1998. In 1999 I revised the lecture into conference papers for the Society for Phenomenology and Existential Philosophy and the American Society for Aesthetics and published it as "Autonomy, Negativity, and Illusory Transgression: Menke's Deconstruction of Adorno's Aesthetics," *Philosophy Today*, SPEP Supplement (1999): 154–68. This essay forms the basis for Chapter 1. Materials for Chapter 5 come from papers I presented to the Critical Theory Roundtable in 2001 and to the American Society for Aesthetics in 2002. Significantly reworked, these papers have appeared as "Feminist Politics and the Culture Industry: Adorno's Critique Revisited," in *Feminist Interpretations of Theodor Adorno*, edited by Renée Heberle (University Park: Pennsylvania State University Press, 2006), pp. 257–76. Later I decided the book should include, as an appendix, a version of my online entry "Theodor Adorno," in *The Stanford Encyclopedia of Philosophy*, edited by Edward N. Zalta (Summer 2003 edition), http://plato.stanford.edu/archives/sum2003/entries/adorno/. Readers who are not well versed in Adorno's writings may want to start with the Appendix, for it provides a succinct overview of his work. I wish to thank the editors and publishers for permission to include revised versions of all three essays here.

Between these bookends occur three chapters I have written since taking up a position at the Institute for Christian Studies (ICS), with cross appointments to the Advanced Degree Faculty at the Toronto School of Theology and to the Graduate Faculty in Philosophy at the University of Toronto. Chapter 2 stems from two papers I presented in 2003 to mark the centennial of Adorno's birth. Together they make up an essay titled "Metaphysics after Auschwitz: Suffering and Hope in Adorno's *Negative Dialectics*," in *Adorno and the Need in Thinking*, edited by the Adorno Research Group at York University, to be published by the University of Toronto Press in 2007. Chapter 3 incorporates a paper read in Montreal at a 2004 conference on Heidegger and Adorno. The essay version will appear as "Truth and Authentication: Heidegger and Adorno in Reverse," in the conference volume *Adorno and Heidegger: Philosophical Questions*, edited by Iain Macdonald and Krzysztof Ziarek, to be published by Stanford University Press in 2007. Upon deciding to turn these materials into a book, I also wrote a new chapter on Horkheimer and Adorno's *Dialectic of Enlightenment*. First completed in July 2005 and presented on several occasions in subsequent months, a slightly different version of Chapter 4 will appear in a book on secularity and globalization edited by James K. A. Smith. A footnote on each chapter-opening page provides more details about that chapter's origins and specific acknowledgments of scholars who commented on earlier drafts. I am especially grateful for sustained discussions with Deborah Cook and Ron Kuipers about the topics of this book.

After publishing *Adorno's Aesthetic Theory* in 1991, I set much of my Adorno scholarship aside to begin a two-volume project on "Cultural Politics and Artistic Truth." A strong impetus toward resumed study came during a three-month visit to Frankfurt in the spring of 2001. There I did new research in the Theodor W. Adorno Archive and participated in Axel Honneth's seminar at the Goethe Universität. I also gave invited lectures at Cambridge University, the University of Edinburgh, the Katholieke Universiteit in Leuven, Belgium, and the Universität Gesamthochschule in Kassel, Germany. I am grateful for gracious hospitality at each of the sponsoring institutions as well as for financial support from the German Academic Exchange Service (DAAD) and from Calvin College, where I taught until 2002.

My writing since then has benefited from simultaneously teaching graduate students at three different schools. The following colleagues deserve special mention for facilitating these interinstitutional arrangements and thereby supporting my own scholarly efforts: Robert Gibbs and James Brown, graduate coordinators and associate chairs in the Department of Philosophy at the University of Toronto; Donald Ainslie, the department chair; and Ansley Tucker, who served as associate academic dean at ICS for two years and became a dear friend. Four students in a guided reading course on "Culture and Economy" read and discussed the entire book manuscript: Benjamin Groenewold, Kristina Jung, Peter Lok, and Tricia Van Dyk. Matt Klaassen, who joined us for the final session, has provided valuable research assistance throughout the project. I want to acknowledge the interest and inspiration of these students and of many others whom I have not named.

Theodor Adorno once recorded a dream in which he refused to abandon his metaphysical hopes because he wanted to awaken together with Gretel, his beloved companion and spouse. Something like Adorno's dream clings to this book on his social philosophy. A year before I decided to write it, our goddaughter Esther Hart and her partner David Roy gave birth to their first and only child. They named her Sophie Marieke. A half year later we learned that at age thirty-six Esther had colorectal cancer. Because the cancer had gone undetected far too long, it had metastasized to several other internal organs. Surgery, chemotherapy, and radiation treatments would follow, temporarily slowing the cancer's growth but not reversing it. Esther wants to live long enough to see her daughter begin school at age four.

Perhaps I would never have met my wife Joyce Alene Recker if she had not moved to Toronto in the early 1970s to take care of a four-year-old girl and her brother when their mother was in the hospital. That little girl was Esther Hart. Three decades later Joyce provides daily care for Esther's daughter Sophie and supports Esther too. Every day Joyce experiences the joy of new life amid the sadness of a loved one's struggle to stay alive. Writing a book on Adorno pales in significance with such heartfelt labors of affection. My life and work in the past few years have been inspired by Joyce's constant compassion, strengthened by Esther's quiet courage, and enlightened by the playful wisdom of Sophie Marieke, their little girl. With gratitude and admiration I dedicate this book to them, sisters across three generations. Death will not defeat them because love keeps us alive.

Abbreviations

Citations of works listed in both English and German use abbreviations derived from the English title. Reference is given first to the English translation and then to the German original, thus: *ND* 153/156. Frequently translations are emended. Dates immediately after titles indicate when the German originals were first published. The Bibliography contains additional works by Adorno and other authors.

BOOKS BY ADORNO

AT   *Aesthetic Theory* (1970), trans., ed., and with a translator's introduction by Robert Hullot-Kentor (Minneapolis: University of Minnesota Press, 1997)/*Ästhetische Theorie*, *Gesammelte Schriften* 7, ed. Gretel Adorno and Rolf Tiedemann, 2d ed. (Frankfurt am Main: Suhrkamp, 1972)

CM   *Critical Models: Interventions and Catchwords* (1963, 1969), trans. Henry W. Pickford (New York: Columbia University Press, 1998)/*Gesammelte Schriften* 10.2 (Frankfurt am Main: Suhrkamp, 1977)

DE   Max Horkheimer and Theodor W. Adorno, *Dialectic of Enlightenment: Philosophical Fragments* (1947), ed. Gunzelin Schmid Noerr, trans. Edmund Jephcott (Stanford, Calif.: Stanford University Press, 2002)/*Dialektik der Aufklärung*, in

Max Horkheimer, *Gesammelte Schriften, Band 5: "Dialektik der Aufklärung" und Schriften 1940–1950*, ed. Gunzelin Schmid Noerr (Frankfurt am Main: Fischer Taschenbuch, 1987)

MM   *Minima Moralia: Reflections from Damaged Life* (1951), trans. E. F. N. Jephcott (London: NLB, 1974)/*Minima Moralia: Reflexionen aus dem beschädigten Leben, Gesammelte Schriften* 4, 2d ed. (Frankfurt am Main: Suhrkamp, 1996)

ND   *Negative Dialectics* (1966), trans. E. B. Ashton (New York: Seabury Press, 1973)/*Negative Dialektik, Gesammelte Schriften* 6 (Frankfurt am Main: Suhrkamp, 1973)

PMP   *Problems of Moral Philosophy* (1963), ed. Thomas Schröder, trans. Rodney Livingstone (Stanford, Calif.: Stanford University Press, 2000)/*Probleme der Moralphilosophie* (1963), ed. Thomas Schröder (Frankfurt am Main: Suhrkamp, 1996)

OTHER AUTHORS

EG   Albrecht Wellmer, *Endgames: The Irreconcilable Nature of Modernity; Essays and Lectures*, trans. David Midgley (Cambridge, Mass.: MIT Press, 1998)/*Endspiele: Die unversöhnliche Moderne; Essays und Vorträge* (Frankfurt am Main: Suhrkamp, 1993)

PDM   Jürgen Habermas, *The Philosophical Discourse of Modernity: Twelve Lectures*, trans. Frederick Lawrence (Cambridge, Mass.: MIT Press, 1987)/*Der philosophische Diskurs der Moderne: Zwölf Vorlesungen* (Frankfurt am Main: Suhrkamp, 1985)

PP   Jürgen Habermas, *Philosophical-Political Profiles*, trans. Frederick G. Lawrence (Cambridge, Mass.: MIT Press, 1983)/*Philosophisch-politische Profile* (Frankfurt am Main: Suhrkamp, 1971)

SZ   Martin Heidegger, *Sein und Zeit* (1927), 15th ed. (Tübingen: Max Niemeyer, 1979)

TCA   Jürgen Habermas, *The Theory of Communicative Action*, trans. Thomas McCarthy, 2 vols. (Boston: Beacon Press, 1984, 1987)/*Theorie des kommunikativen Handelns* (Frankfurt am Main: Suhrkamp, 1981)

Introduction

Thinking Otherwise

> The only philosophy which can be responsibly practised in face of despair [would be] the attempt to contemplate all things as they would present themselves from the standpoint of redemption.
>
> Adorno, *Minima Moralia*[1]

I remember well the month when I began to read Theodor W. Adorno's *Ästhetische Theorie*. It was May 1977, during a lovely spring in Toronto. Joyce and I were house-sitting for a professor of political philosophy. In the quiet of someone else's study, surrounded by books that were not my own, I began to read Adorno's impenetrable, compelling, evocative German prose. Some days I made little headway. Other days I found myself swept along by the drama of the text, yet unable to tell anyone else where I had been or what I had learned. Gradually, however, I began to glimpse the submerged dialectical structures that sustain Adorno's thought.

My reading notes from 1977 show that I experienced *Ästhetische Theorie* as an array of potentially interconnected fragments held

---

An earlier version of section 1 in this Introduction was presented at the symposium "Adorno – Cultural Theory, Political Thought and Social Change," hosted in November 2003 by the Goethe Institut in Toronto to mark the centennial of Adorno's birth. I wish to thank Doina Popescu and Arpad Sölter of the Goethe Institut for organizing this occasion, and my fellow panelists for providing a lively discussion: Ian Balfour, Rebecca Comay, Lydia Goehr, Willi Goetschel, Richard Leppert, and Asha Varadharajan.

[1] *MM* §153, p. 247/283.

together by a provocative social vision. Although I had little understanding then of Adorno's emphasis on parataxis and constellations, I read bits and pieces, relying on intuitions about their relevance for my own concerns: first several pages on modern art (*AT* 33–45/56–74), next the chapter on truth content (*AT* 118–36/179–205), then some pages on progress (*AT* 190–9/284–96) and on society (*AT* 225–35/334–48, 256–61/380–7), and finally a few pages from the "Paralipomena" (*AT* 282–5/419–24). In the months surrounding this first exposure, my reading ranged through *Philosophy of New Music*, *Minima Moralia*, and *Dialectic of Enlightenment* as well as some German secondary literature, seeking further illuminations of Adorno's text. This is how my reading of his unfinished *Hauptwerk* continued for several years – fragmentary, contextual, and out of sequence. Not until Christian Lenhardt sent me an autographed copy of his newly published translation in 1984 did I read the book continuously from beginning to end, in English now, with the German text alongside. By then the main lines of my own interpretation were firmly established.

A few months after the first partial reading, Joyce and I immigrated to West Berlin, where over several years I researched and wrote my doctoral dissertation. We arrived there in the fateful fall of 1977. Even before we could catch nuances of newscasts and conversations, we felt the tension tingeing the famous "Berliner Luft." We had moved from spring to autumn, and my reading of Adorno took on a more somber cast. Awakened frequently in the dead of night by the ominous rumble of American tanks maneuvering down Teltower Damm, I soon lost the can-do optimism that came with growing up white, male, and middle class in the United States at the height of its Cold War empire. By the time we had lived in Berlin for three years, at the epicenter of geopolitical conflicts, Adorno's harshest criticisms of capitalist society seemed none too harsh. The "sadness" and "disappointment" lodged at the heart of Adorno's negative dialectic, as Max Pensky puts it, resonated with the world around.[2]

---

[2] Max Pensky, "Editor's Introduction: Adorno's Actuality," in *The Actuality of Adorno: Critical Essays on Adorno and the Postmodern*, ed. Max Pensky (Albany: State University of New York Press, 1997), pp. 1–21.

Introduction: Thinking Otherwise

Yet the book that eventually emerged from this experience received quite a different response.[3] At a book session organized by the Society for Phenomenology and Existential Philosophy in October 1992, one commentator regarded *Adorno's Aesthetic Theory* as a domesticating reconstruction of Adorno's thought. The other considered it a depoliticizing critique. Surprised by this, I began to wonder about an apparent gap between the book's presentation of Adorno and the lived hermeneutical experience from which this presentation emerged.

In retrospect, I can see that the gap arose from my intending the book to address a broad audience. This audience included analytic philosophers who might otherwise pay Adorno's philosophy little notice, university students around the world who needed a rigorous but accessible account of Adorno's contributions, and cultural workers who wanted to challenge the logic of consumer capitalism. Like most of my work since then, the book did not restrict its audience to continental philosophers, Critical Theorists, and Adorno cognoscenti, even though it invited engagement from scholars who had particularly strong reasons to read Adorno in the 1990s.

Nor did the book set out to provide either the passionate meditations or the anti-imperialist fireworks that my commentators would have preferred. Instead, it attempted to give an immanent critique with metacritical intent, focusing on the theme of artistic truth in Adorno's aesthetics. One commentator wanted the critique to be less immanent, and the other wanted the immanence of my critique to be less dispassionate. Neither expectation struck me as what was needed at the time. Yet by attempting an immanent critique with metacritical intent, I had left underdeveloped the critical passions and political relevance of Adorno's work.

Today I am in a position to close the gap. This is so in part because Adorno's thought has become better known in the English-speaking world, although it still does not receive the intensity and breadth of study it deserves. The gap can also be closed because the issues at stake in reading Adorno have become more transparent, not only for me but also for many others who address his work. It is not hard to see now that the future of a globalizing society is at stake. This is especially

---

[3] Lambert Zuidervaart, *Adorno's Aesthetic Theory: The Redemption of Illusion* (Cambridge, Mass.: MIT Press, 1991).

clear if one considers the development of Critical Theory, in which the question of philosophy's social vision has always been central. A dramatic shift has occurred in this development, such that one can recognize two different visions of socially engaged philosophy.

I.1 WOZU NOCH PHILOSOPHIE?

A first approximation of two different visions comes from comparing essays by Adorno and by Jürgen Habermas, both titled "Wozu noch Philosophie." Adorno's essay, translated as "Why Still Philosophy," was broadcast as a radio lecture on January 2, 1962 (*CM* 1–17/459–73).[4] It comes from the time when he was writing *Negative Dialectics*. The essay objects to the formalism of much professional philosophy, and it criticizes other schools of thought for ignoring societal mediation: logical positivism, for ignoring the mediation of facts, and Heideggerian ontology, for ignoring the mediation of concepts.

According to Adorno, such immanent criticism of other philosophies has a larger societal purpose. It aims to expose the "unfreedom and oppression" at work in contemporary society (*CM* 10/465). It also aims to "catch a glimpse" (*CM* 15/471) of a world where they would end. Adorno does not hesitate to use strong language when he states philosophy's task. He speaks of "suffering," "salvation," and "hope":

> The undiminished persistence of suffering, fear, and menace necessitates that the thought that cannot be realized should not be discarded. After having missed its opportunity, philosophy must come to know, without any mitigation, why the world – which could be paradise here and now – can become hell itself tomorrow. (*CM* 14/470)

> Only a thinking ... that acknowledges its lack of function and power can perhaps catch a glimpse of an order of the possible and the nonexistent, where human beings and things each would be in their rightful place. (*CM* 15/471)

> History promises no salvation and offers the possibility of hope only to the concept whose movement follows history's path to the very extreme. (*CM* 17/473)

---

[4] The original version of this essay dates back to a talk Adorno gave in 1955 for a study group in the Frankfurt Student Union, according to Stefan Müller-Doohm, *Adorno: A Biography*, trans. Rodney Livingstone (Cambridge: Polity Press, 2005), p. 416.

Introduction: Thinking Otherwise 5

For Adorno, the pursuit of this vision requires a philosophy whose experience is unrestricted (*CM* 17/473).

Habermas's essay, translated as "Does Philosophy Still Have a Purpose?" also began as a radio lecture (*PP* 1–19/11–36). Broadcast on January 4, 1971, it forms the lead essay in a volume dedicated to Adorno's memory: "In Erinnerung an Theodor W. Adorno," reads the dedication in Habermas's *Philosophisch-politische Profile*. The book's English translation omits this dedication, however, just as Adorno's utopian passion had already vanished from Habermas's lecture. When Adorno asked "Wozu noch Philosophie?" he wondered what philosophy could contribute to transforming society as a whole. This is no longer Habermas's question.

One can detect the shift from how Habermas cites Adorno. The essay opens by quoting a passage where Adorno says philosophy must no longer consider itself in control of "the absolute," yet it must retain "the emphatic concept of truth." Habermas ends the quotation with Adorno's sentence "This contradiction is its element" (*CM* 7/461; quoted in *PP* 1/11). Significantly, Habermas omits the very next sentence in Adorno's text, a sentence that is equally important: "It [i.e., this contradiction] defines philosophy as negative," Adorno writes (*CM* 7/461). Habermas does not speak about negativity, about the negativity of suffering, say, or the negativity of a societal totality that needlessly produces and prolongs suffering. Resting content with a conception of nonabsolute truth, he becomes nearly elegiac about the distance of his vision from Adorno's. Adorno's death marks the end of a "great tradition" of German philosophy, Habermas writes, and with it a "style of thought bound to individual erudition and personal testimony" (*PP* 2/12). Quite rightly, I think, he wonders whether, in catching up with modernization in other Western countries, German philosophy itself will "fade away in the graveyard of a spirit that can no longer affirm and realize itself as absolute" (*PP* 9/22).

Habermas does not wish philosophy to fade away. Yet his essay limits philosophy's contemporary tasks to a "substantive critique of science" (*PP* 14/30) – a critique of *science*, not of *society as a whole*. Although the critique is supposed to be "substantive," the specific tasks are notably formal in their description: "to criticize the objectivist self-understanding of the sciences," "to deal ... with basic issues of a methodology of the social sciences," and to clarify connections between "the logic of research and technological development," on the one hand, and "the logic of consensus-forming communication," on the other (*PP* 16/33). Habermas relegates questions of suffering and hope to religion, which itself has become impotent in "industrially advanced societies" (*PP* 18/35).[5] One finds no sign of Adorno's emphasis on philosophical experience, and little trace of his desire to expose the negativity of society as a whole. Whether, in abandoning Adorno's struggle with "the absolute," Habermas has also lost "the emphatic concept of truth" remains an open question.

Perhaps this comparison suffices to show a dramatic shift within Critical Theory less than two years after Adorno's death in August 1969. The question I want to pose is whether the shift matters. My answer is that it does, in three respects. First, it supports serious misinterpretations of Adorno's thought. Second, it blunts the political edge of Critical Theory. Third, it results in a truncated vision of philosophy at a time when passion, not simply precision, is required. This book attempts to retrieve some of Adorno's passion without neglecting his dialectical precision. It pursues what I call a "critical retrieval."

I aim to retrieve crucial insights in Adorno's social philosophy. By "social philosophy" I do not mean a subdiscipline of philosophy that can be neatly arrayed alongside other subdisciplines such as epistemology, ethics, and aesthetics. Rather I mean the entirety of philosophy as it addresses the challenges and the prospects of society as a whole. As Adorno demonstrated, such a philosophy is inherently interdisciplinary. It interacts with other disciplines in order to undertake a dialectical critique of society, and it necessarily crosses the boundaries of epistemology, ethics, and aesthetics, even as it

---

[5] This move occurs in an underdeveloped reflection on how contemporary philosophy confronts a "collapse of religious consciousness." The term in German, *Zerfall*, is the same one Adorno sometimes uses to characterize the demise of metaphysical consciousness. Habermas regards this collapse as a challenge for philosophy because philosophical life-interpretations among the cultured elite traditionally "depended precisely on coexistence with a widely influential religion," but philosophy itself was unable to master "the meaninglessness of the negativity of the risks built into life – in a way that had been possible for the religious hope in salvation [*die Erwartung des religiösen Heils*]" (*PP* 17–18/35).

addresses topics within each of these fields. To regard all things "as they would present themselves from the standpoint of redemption," as Adorno urges, is to surpass the limits imposed by overly professionalized and hyperspecialized philosophies. It is to engage in the risky and provocative project of thinking otherwise.

I.2 GOING AFTER ADORNO

In the effort to retrieve Adorno's social philosophy I am not alone. A number of books have appeared in recent years that try to reclaim Adorno's insights from Habermasian neglect. Their authors share my desire to do philosophy "after Adorno," in a double sense. First, we wish to carry forward crucial insights in Adorno's social philosophy. Second, we try to do this by "going after" Adorno's successors such as Habermas, refuting their criticisms of Adorno and reclaiming the Adornian insights they overlook or reject.

Yet there is a third manner of doing philosophy after Adorno that deserves greater effort than it has received thus far. This is the project of acknowledging valid Habermasian objections and, in light of them, providing a redemptive critique of Adorno. It is not enough, in my view, to defend Adorno against misinterpretations, to reject inadequate criticisms, and to promote the concerns and claims his successors neglect. One also needs to address legitimate Habermasian criticisms of Adorno and suggest viable alternatives. Otherwise the return to Adorno will not be a fully critical retrieval. Accordingly, this book takes up Habermasian criticisms of Adorno, explores whether these are valid, and offers alternatives that, while inspired by Adorno's social philosophy, also avoid its problems. Through such critical retrieval, I propose new directions for a social philosophy "after Adorno": one that, being indebted to Adorno, also "goes after" him, but only by "going after" Adorno's loyal critics.

In this connection, let me comment briefly on the books to which my project is most closely related and most strongly indebted.[6]

---

[6] Here I shall not discuss other recent books that question the Habermasian reception of Adorno's work but with greater emphasis on aesthetics or epistemology. See, for example, Yvonne Sherratt, *Adorno's Positive Dialectic* (Cambridge: Cambridge University Press, 2002); Pieter Duvenage, *Habermas and Aesthetics: The Limits of Communicative Reason* (Malden, Mass.: Polity Press, 2003); Brian O'Connor, *Adorno's*

Simon Jarvis has provided the best book-length introduction to Adorno's thought to date.[7] Like Jarvis, I consider all of Adorno's work to be interconnected, as my book's Appendix demonstrates in short compass. But Jarvis too readily accepts Habermasian criticisms of Adorno's "metaphysics," and he leaves Adorno's aesthetics intact. If Adorno's writings on art and the culture industry belong to a larger project of social philosophy, then his central aesthetic claims also need to be reexamined in a social-philosophical context. This I attempt to do in Chapters 1 and 5, in conjunction with the intervening chapters. Specifically, in Chapter 1 I criticize Christoph Menke's reconstruction of Adorno's socially engaged aesthetics and suggest an alternative to both Menke and Adorno on the autonomy of art.

Jarvis's tendency to accept the postmetaphysical turn is challenged by Jay Bernstein.[8] Taking issue with Habermas's discourse ethics, Bernstein derives a substantial "modernist ethics" from Adorno's thought. I share Bernstein's aim of retrieving Adorno's work from Habermasian postmetaphysical criticisms. The key to Bernstein's retrieval lies in a theory of "the complex concept" that opposes Albrecht Wellmer's criticism of Adorno as having engaged in a critique of conceptual knowing as such. This theory allows Bernstein to read Adorno as offering an ethical alternative to the "disenchantment" and "nihilism" that accompany modernization. Although largely in agreement with Bernstein's interpretation of Adorno, Chapter 2 identifies two problems that motivate Wellmer's criticisms and that Bernstein tends to overlook. It also defends Adorno's "metaphysical experience" against Wellmer's postmetaphysical criticisms. By offering an alternative to both Adorno and Wellmer in the next two chapters, I also elaborate questions posed in my review of Bernstein's book.[9]

Negative Dialectic: Philosophy and the Possibility of Critical Rationality (Cambridge, Mass.: MIT Press, 2004). A crucial opening salvo in the struggle to reclaim Adorno from the Habermasians is Robert Hullot-Kentor, "Back to Adorno," *Telos*, no. 81 (Fall 1989): 5–29.
[7] Simon Jarvis, *Adorno: A Critical Introduction* (New York: Routledge, 1998).
[8] J. M. Bernstein, *Adorno: Disenchantment and Ethics* (Cambridge: Cambridge University Press, 2001).
[9] See *Constellations* 10 (2003): 280–3.

Introduction: Thinking Otherwise

Like Bernstein, Martin Morris resists the deaestheticization of social philosophy in Habermasian Critical Theory.[10] In an effort to recover Adorno's insights for political and ethical thought, Morris discovers potential for democratic communication where Habermas sees only aesthetic gestures. Although sympathetic to Morris's project, I think his proposed "politics of the mimetic shudder" fails to recognize the problems in Adorno's appeal to "emphatic experience" and in his idea of truth. I take up these problems in Chapter 3, with a view to developing a normative theory of democratic truth telling that goes beyond Habermas's political philosophy.

Such a theory would need to consider the structure and dynamic of late capitalist society as a whole. Addressing this topic, Deborah Cook regards Habermas's account of the "colonization of the lifeworld" as too sanguine, and she defends Adorno's critique of "domination" as diagnostically more astute and politically more progressive.[11] She also suggests that, in later writings on globalization, Habermas might in fact be returning to the Adornian fold. That is to say, the later Habermas might actually recognize the pervasiveness of economic exploitation under conditions of globalization. Although I find Cook's defense of Adorno instructive, she overlooks Habermas's normative questions concerning societal rationalization. Correlatively, she fails to challenge Adorno's insufficiently differentiated idea of domination. Chapter 4 wrests Adorno's critique of capitalism from the grip of Habermas's powerful misinterpretation, in order to point toward a normative theory of globalization. As sketched in Chapter 4, this theory takes Cook's critique of economic globalization in a more hopeful direction.

The politics of culture is central for such a critique, both in Adorno's social philosophy and in contemporary attempts to theorize globalization. Unlike critics who fault Adorno for an alleged lack of political engagement, Espen Hammer argues that Adorno "was one of the most politically acute thinkers of the twentieth century."[12] Acknowledging that Adorno never developed a political theory,

---

[10] Martin Morris, *Rethinking the Communicative Turn: Adorno, Habermas, and the Problem of Communicative Freedom* (Albany: State University of New York Press, 2001).
[11] Deborah Cook, *Adorno, Habermas, and the Search for a Rational Society* (New York: Routledge, 2004).
[12] Espen Hammer, *Adorno and the Political* (New York: Routledge, 2005), p. 1.

Hammer says that Adorno's contributions to political thought lie in his specific critical interventions. I agree with Hammer that Adorno's work as an educator and public intellectual made significant contributions to progressive politics. Yet the absence of a full-fledged political theory strikes me as a significant deficit, especially because Adorno's social theory casts doubt on all collective political struggles for liberation. I take up this topic in Chapter 5, where I explore the implications of Adorno's critique of the culture industry for a revitalized feminist politics. Unlike Hammer, I do not examine how Adorno thought, or would have thought, about so-called feminist issues.[13] Rather, I ask whether certain insights in Adorno's critique of the cultural economy, if released from the blinders of an inadequate political theory, are relevant for contemporary feminist politics in ways that Habermasian Critical Theory is not.[14] The broader implications of my disagreements with these fellow Adornians emerge in Chapter 6, where I propose a democratic politics of global transformation.

1.3 CRITICAL RETRIEVAL

Following the trajectory of Adorno's own life work, and of my own engagement with it over the years, this exercise in critical retrieval begins with a topic that, in another philosophy, might be considered "merely aesthetic": the autonomy of art. As Chapter 1 demonstrates, this topic is not merely aesthetic, nor is it peripheral to Adorno's social philosophy as a whole. For Adorno's aesthetics employs a complex idea of artistic autonomy. Modern art is the social antithesis of society, he asserts. Because Western society strips art of overt social functions, the best modern art can engage in a determinate negation of society and thereby offer both utopian vision and social critique. Dissatisfied with seemingly exaggerated claims that accompany Adorno's account of artistic autonomy, Christoph Menke, a former student of Albrecht Wellmer, tries to rearticulate Adorno's "aesthetics of

---

[13] Ibid., p. 171.
[14] By pursuing this question, Chapter 5 both elaborates and modifies my earlier criticisms of Adorno's cultural politics for devaluing "actual struggles for political liberation" and thereby ending up as "a merely cultural politics." Zuidervaart, *Adorno's Aesthetic Theory*, p. 149.

Introduction: Thinking Otherwise

negativity." After summarizing Menke's Derridean reconstruction, I argue that he reduces Adorno's complex idea of artistic autonomy in two respects. He reduces the autonomy of art to the autonomy of the aesthetic dimension, and he reduces the autonomy of the aesthetic to the supposed negativity of aesthetic experiences. This double reduction renders art politically irrelevant. As a result, Menke forgets why Adorno considers autonomy so crucial, namely, as a way to preserve critical and utopian capacities in a late capitalist society. Nevertheless, Menke has called attention to problematic aspects of Adorno's account of artistic autonomy. Chapter 1 concludes by suggesting an alternative account, to be elaborated in subsequent chapters. While avoiding Menke's double reduction, my alternative transforms Adorno's paradoxical modernism.

The issues raised by Adorno's aesthetics are not simply aesthetic concerns. They are societal matters that have ethical, political, and epistemological dimensions. Two large topics are at stake. One has to do with how philosophy should respond to societal evil. The other concerns the appropriate philosophical diagnosis of contemporary society and its prospects. Chapters 2–4 pursue these topics in *Negative Dialectics* and *Dialectic of Enlightenment*. Then I revisit Adorno's idea of artistic autonomy in Chapter 5 and demonstrate the relevance of my critical retrieval for progressive politics in Chapter 6.

Chapter 2 summarizes and questions Albrecht Wellmer's postmetaphysical critique of Adorno's meditations on metaphysics. In response, I explicate the Adornian themes of suffering and hope as ones that postmetaphysical philosophy mistakenly neglects. I show that Adorno's emphasis on suffering goes hand in hand with hope for a fundamentally transformed society within which thought itself would be thoroughly transformed. Then I explore two problems in Adorno's thematization of suffering and hope. The first pertains to Adorno's privileging of "philosophical experience." I argue that this makes the philosophical interpretation of suffering self-authenticating. The second problem concerns what I call Adorno's "objectification of transformative hope." For Adorno, hope arises from objects that have resisted the principles of identification and exchange. Because nonidentical objects are an inadequate basis for societal transformation, Adorno's hope seems both crucial and ill-supported. Habermasian criticisms such as Wellmer's try to avoid these problems but do so at the price of becoming postmetaphysical. What is required instead, I claim, is a social philosophy for which suffering is real and for which transformative hope is not misplaced.

Chapter 3 continues my critical retrieval of Adorno's emphasis on experience. There I claim that, in their conceptions of how truth is authenticated, the dialectical extremes of twentieth-century German philosophy touch. Whereas Martin Heidegger says the process occurs in the "authenticity" of human existence (Dasein), Adorno locates the authentication of truth in "emphatic experience." Portraying existential authenticity and emphatic experience as each other's reverse image, I argue that neither suffices to authenticate truth. Then I offer an account of public authentication that draws out the social significance of these dialectical extremes. Both Adorno and Heidegger recognize that philosophical conceptions of truth have far-reaching implications for the future of society and of human existence. I suggest that their philosophies challenge us to discover how truth can be borne out in ways that, while authentic and emphatic, are also democratic.

Whereas Chapter 3 partially recuperates Adorno's emphasis on experience, Chapter 4 returns to his objectification of transformative hope, the second tendency problematized in Chapter 2. I do this by linking central claims in *Dialectic of Enlightenment* with contemporary debates about globalization. The chapter begins with Habermas's well-known objections to this seminal text. Habermas accuses Adorno of totalizing reification, thereby undermining Critical Theory's normative foundations, doing injustice to modern differentiation, and seeking an aesthetic escape. I show that Habermas's objections deploy a serious and sustained misreading of the "remembrance of nature" in Adorno's thought. Adorno's "remembrance" is not an aesthetic gesture. Rather, it is a conceptual process of critical self-reflection.

Hence, the theoretical moves that lead to Adorno's objectification of transformative hope cannot be explained along Habermasian lines. They have to do, first, with the scope and imbrication of domination and, second, with the connections between economic exploitation and cultural differentiation. On the first issue, I distinguish more clearly than Adorno does among three modes of domination: control, repression, and exploitation. I agree with Adorno that all three modes characterize modern Western society and that understanding their interlinkage is crucial for a transformative social theory. But I also propose a distinct form of normative critique for each mode of domination. On the second issue, I put forward, more forthrightly than either Adorno or Habermas does, a normative critique of the economic system. This critique would envision a differential transformation of society as a whole. By "differential transformation" I mean a process of significant change that occurs at differing levels, across various structural interfaces, and with respect to distinct societal principles. In this process, societal principles such as justice and solidarity will be mutually complementary, not only in the economy, where they are often violated or ignored, but also in the state, in civil society, and, indeed, in society as a whole. Just as Chapter 3 redirects Adorno's emphasis on emphatic experience into a call for public authentication, then, so Chapter 4 turns Adorno's speculation about nonidentical objects into a vision of society's differential transformation.

Returning to the discussion of autonomy begun in Chapter 1, Chapter 5 tests the relevance of my critical retrieval of Adorno's social philosophy for feminist cultural politics. Chapter 5 examines Adorno's critique of the culture industry, the contribution for which he is best known. I argue that Adorno's critique identifies crucial issues that contemporary feminism needs to address. After identifying a tension within feminism concerning the idea of artistic autonomy, the chapter examines the role this idea plays in Adorno's critique of the culture industry. Adorno's innovative and productive move is to interpret the modern Western notion of artistic autonomy by updating Marx's dialectic of the commodity. This allows Adorno to connect the internal and self-critical independence of the authentic work of art (internal artistic autonomy) with high art's relative independence from the political and economic system (societal artistic autonomy) and also with the autonomy of the self as a political and moral agent (personal autonomy). Adorno criticizes the culture industry for undermining all three types of autonomy.

But there are problems with how Adorno links internal and societal autonomy, as I demonstrate from two passages in *Dialectic of Enlightenment*. Adorno's critique of the culture industry goes wrong when it portrays cultural goods as *no more than* hypercommodities.

I respond by proposing a reconfigured notion of artistic autonomy. As Adorno's linking of societal and internal autonomy suggests, countereconomic and counterpolitical spaces must be developed, and they must foster intrinsically worthwhile cultural practices, including art practices. Unlike Adorno, I see potential for such spaces and practices in the third sector of a mixed economy and in contemporary public spheres. An account of artistic autonomy developed along these lines would correct Adorno's dialectic between hypercommercialized culture and authentic artworks. It would also dissolve a conflict in feminism between strategically appealing to artistic autonomy and theoretically rejecting it.

Hence, my account of public authentication and differential transformation, developed through immanent criticism of Adorno and his critics, has productive cultural political implications. In fact, Adorno's critique of the culture industry provides a test case for the critical retrieval this book proposes. On the one hand, Adorno's critique demonstrates the promise of a social philosophy oriented toward the transformation of society as a whole. Such a philosophy uncovers connections and contradictions where others do not dare or care to look. On the other hand, the limitations to Adorno's critique of the culture industry are themselves connected to problems in his philosophy as a whole. His pitting internally autonomous artworks against the culture industry supports his attempt to secure sources of "emphatic experience." And his treating culture-industrial goods as no more than hypercommodities manifests the shortcomings to his objectification of transformative hope.

So too, my own reconfiguration of "artistic autonomy" points toward the ideas of public authentication and differential transformation that this book offers in response to these problems. The debate over artistic autonomy concerns nothing less than the question, What configuration of art practices within which institutions and societal structures would enable people to flourish in their aesthetic pursuits? And, more broadly, which configuration would be more conducive to human freedom? Further, if the current configuration is deeply flawed, as, like Adorno, I believe it is, what simultaneous and mutually reinforcing changes would be required – both at the differing levels of social institutions, cultural practices, and interpersonal relations and at the structural interfaces among economy, polity, and culture – in order for societal principles such as justice and solidarity better to hold sway? In other words, what differential transformation would take society, and the arts within society, in a better direction? Authentic works of art may not be the monads of society that Adorno took them to be, but the debate over artistic autonomy, when connected with Adorno's critique of the culture industry, can serve as a catalyst for contemporary projects of critical retrieval.

Chapter 6 extends the claims in previous chapters by exploring the broader political implications of Adorno's social philosophy. I take issue with a tendency among Adornian critical theorists to reduce his politics to a personal ethics. To develop appropriate responses to globalization requires a politics informed by social ethics, not one reduced to personal ethics. It also requires a political theory that, unlike Adorno's, articulates collective agencies and normative principles. To those ends Chapter 6 revisits Adorno's reflections on theory and practice, calls for a robust theory of social democracy, and discusses the prospects for a global civil society. The chapter also proposes a vision of societal disclosure that could provide orientation for a global social ethic.

I conclude by considering possible Adornian objections to my emphasis on collectivity and normativity in a global context. I grant that Adorno's negative dialectic does not lend itself to this emphasis. Yet I argue that an emphasis on collectivity and normativity is necessary to take seriously the "new categorical imperative" that Adorno proposes for an ethics after Auschwitz. For the demand to resist any recurrence of such societal evil cannot be a moral imperative unless it can be met. And it can be met only if our resistance has ethical resources and political agencies that are both collective and normative. Accordingly, Adorno's politics should not be reduced to a stance of personal resistance, no more than his social philosophy should be reduced to either ethics or aesthetics. Rather, his ethics of resistance should be rearticulated in a democratic politics of global transformation. For such a politics, Adorno's passion and precision hold promise. For a social philosophy that would not capitulate "in face of despair," critically retrieving Adorno's insights is an urgent matter.

1

Transgression or Transformation

> Insofar as a social function can be predicated for artworks, it is their functionlessness. Through their difference from a bewitched reality, they embody negatively a position in which what is would find its rightful place, its own.
>
> Adorno, *Aesthetic Theory*[1]

The account of autonomy in Adorno's aesthetics hinges on the dysfunctional social function of authentic artworks. This account marks a theoretical passageway both to his critique of the culture industry and to his qualified defense of modern art. Neither the critique nor the defense makes sense in isolation from larger themes in Adorno's social philosophy. Let me show this by briefly summarizing his approaches to the culture industry and to modern art.

Adorno's critique of the culture industry has three levels, as Deborah Cook has demonstrated:[2] an aesthetic thesis about the formulaic character common to products of the culture industry, an economic thesis about the commodification of culture, and a psychological thesis

---

I wish to thank Zenon Bankowski, Andrew Benjamin, Jay Bernstein, James Bohman, Maeve Cooke, Gregg Horowitz, Tom Huhn, Russell Keat, Christoph Menke, Heinz Paetzold, Max Pensky, Karla Schultz, Calvin Seerveld, and my former colleagues in the Philosophy Department at Calvin College for their comments on earlier drafts of this chapter.

[1] *AT* 227/336–7.
[2] Deborah Cook, *The Culture Industry Revisited: Theodor W. Adorno on Mass Culture* (Lanham, Md.: Rowman & Littlefield, 1996).

about narcissistic tendencies in late capitalist societies. In my view, these levels interconnect through the strong links Adorno forges among three different concepts of autonomy: (1) the internal and self-critical independence of the authentic work of art; (2) the relative independence of (some of) high culture from the economic system; and (3) the autonomy of the self as a political and moral agent. He criticizes the culture industry for undermining all three types of autonomy. And he portrays authentic works of modern art as providing crucial resistance to pressures toward cultural commodification and social narcissism.

The account of autonomy in Adorno's qualified defense of modern art goes roughly like this.[3] Art has become independent from other institutions in capitalist society. Because this independence depends on political, economic, and other developments, however, the independence of art is relative to society as a whole. Art's autonomy is a matter of relative independence. Increasingly this relative independence has become tied to the production and reception of artworks whose "purposes" are internal to the works themselves. Such works, when they are authentic, are social monads. Their internal processes express and challenge those of society as a whole. Being relatively independent, authentic works of modern art can enact a critique of society by engaging in self-criticism. They are like self-inflicted wounds on the body of society. They force people to face society's deeper illness – its divisions of labor, class conflict, and "exchange principle" – and they point toward remedies that art itself cannot prescribe.

Adorno's account of artistic autonomy has come in for a number of criticisms. Peter Bürger faults him for ignoring how art itself, in its avant-garde movements, has attacked the principle of autonomy.[4] Sabine Wilke charges Adorno with using his principle of autonomy in a gender-biased way to ignore or discredit the work of women artists.[5] I have criticized Adorno for making autonomy a

---

[3] Adorno's account of autonomy is, of course, considerably more complicated than a one-paragraph summary can suggest. See especially the first and last chapters of his *Aesthetic Theory*, AT 1–15/9–31 and AT 225–61/334–87.
[4] Peter Bürger, *Theory of the Avant-Garde*, trans. Michael Shaw, with a foreword by Jochen Schulte-Sass (Minneapolis: University of Minnesota Press, 1984).
[5] Sabine Wilke and Heidi Schlipphacke, "Construction of a Gendered Subject: A Feminist Reading of Adorno's *Aesthetic Theory*," in *The Semblance of Subjectivity: Essays*

precondition of truth in art, thereby discounting the truth potential of folk art, mass-mediated art, and any art prior to the solidification of fine art as a social institution in eighteenth-century Europe.[6] Indeed, Adorno's emphasis on the autonomy of art marks a contemporary site of contention. Scholars in feminism, postcolonial theory, and cultural studies criticize it as an antidemocratic vestige of Eurocentric philosophy, while poststructuralists and postanalytic philosophers dismiss it as an antiquated thesis of modernist aesthetics. In response, more sympathetic readers try to rescue Adorno's account by demonstrating its compatibility with the concerns of contemporary scholarship.

Unfortunately, these rescue efforts often fail in two respects: they jettison insights in Adorno's writings that could counter theoretical deficiencies in contemporary scholarship, and they intensify practical conundrums in the arts that Adorno could not adequately address. This chapter takes up both problems. It does so by commenting on a partially deconstructive attempt to reclaim Adorno's account – Christoph Menke's use of Jacques Derrida to rearticulate Adorno's "aesthetics of negativity."[7] After summarizing the argument in Menke's *The Sovereignty of Art*, I take issue with his deconstructive reading of Adorno. I then conclude by sketching what a viable account of autonomy might require. By the chapter's end, I hope to have given some persuasive reasons not to reconstruct Adorno's aesthetics along Derridean lines.

1.1 MENKE'S DERRIDEAN RECONSTRUCTION

Christoph Menke tries to reclaim two strands of modern aesthetics. According to the first strand, aesthetic experience is *autonomous*,

---

in *Adorno's Aesthetic Theory*, ed. Tom Huhn and Lambert Zuidervaart (Cambridge, Mass.: MIT Press, 1997), pp. 287–308.

[6] Lambert Zuidervaart, *Adorno's Aesthetic Theory: The Redemption of Illusion* (Cambridge, Mass.: MIT Press, 1991), pp. 217–47.

[7] Christoph Menke, *The Sovereignty of Art: Aesthetic Negativity in Adorno and Derrida*, trans. Neil Solomon (Cambridge, Mass.: MIT Press, 1998). Cited in text by page number. This is a translation of *Die Souveränität der Kunst: Ästhetische Erfahrung nach Adorno und Derrida* (Frankfurt am Main: Suhrkamp, 1991), itself a slightly revised version of an earlier book by the same title (Frankfurt am Main: Athenäum Verlag, 1988).

adhering to its own internal logic and having its own place alongside other discourses and modes of experience. When this strand dominates, art appears isolated and irrelevant. According to the second strand, aesthetic experience is *sovereign*, exceeding its own internal logic, disrupting all other discourses, and thereby providing "an experientially enacted critique of reason." When this strand dominates in aesthetics, art gets saddled with metaphysical or social-critical burdens it cannot carry. Menke seeks to weave the two strands together "in a logically consistent and comprehensive manner" that prevents their unraveling into either the isolating or the overburdening of art. In this he claims to emulate Adorno, who, refusing to sacrifice either strand, granted both of them "full expression ... in all their mutual tension" (p. viii).

Aesthetic Negativity

In part I (chapters 1–4) Menke explains the key to this project as "aesthetic negativity," a concept he derives from Adorno and refines with the help of Derrida. His thesis is that aesthetic negativity constitutes "the distinction between the aesthetic and the nonaesthetic." By aesthetic negativity he means the negative relationship in which art and artworks stand "to everything that is not art." Art's autonomy – its distinctiveness and uniqueness – consists in art's being "contradiction, rejection, negation" (p. 3). By describing autonomy as negativity, Menke hopes to recapture "the twofold definition of modern art in Adorno, of art as both one of several autonomous discourses and a sovereign subversion of the rationality of all discourses" (p. xi).

Yet Menke argues that Adorno too frequently conflates aesthetic negativity and social critique. One can avoid this conflation, he says, if one defines aesthetic negativity semiotically, "in terms of the use and understanding of signs" (p. xii). At the same time, one can retain the distinctiveness of aesthetic experience by emphasizing, in Kantian fashion, that aesthetic pleasure arises "not in direct confrontation with an object ... but in our reflective recourse ... to the process of experiencing the object." More precisely, aesthetic pleasure arises from the way in which "diverse, nonaesthetic, automatic acts of recognition" are transformed in aesthetic experience (p. 13).

By contrast, other accounts of aesthetic negativity, which Menke labels either "social-critical" or "purist," reduce aesthetic pleasure to either moral or sensuous pleasure.

According to Menke's aesthetics of negativity, then, aesthetic experience is simultaneously an attempt at understanding and a negation of that attempt. In aesthetic experience the unavoidable effort to understand the artwork is unavoidably subverted. Aesthetic pleasure arises from this "releasing of an unsublatable negativity in the negated, which ... alone renders the negated suitable for aesthetic experience" (p. 25).[8] Menke posits, but does not really argue, that all nonaesthetic "enactments of understanding" are automatic: they rely on conventions or rules to identify an object. Only aesthetic enactments are nonautomatic, for they rely on no conventions or rules and neither seek nor achieve "an identificatory end." The *process* is crucial in aesthetic experience, not the result, and this process "is also the subversion of any understanding-based identification of the object." Aesthetic experience negates the "automatic repetition" that characterizes nonaesthetic experience (pp. 31–2). In semiotic terms, *automatic* understanding aims to establish a meaningful link between signifiers selected from materials (sounds, gestures, marks, etc.) and the signifieds. In *aesthetically enacted* understanding, by contrast, the signifier itself interminably vacillates between the poles of material and meaning. Aesthetic experience is "the processual enactment" of this vacillation (p. 46): "The aesthetic signifier is nothing more than this interminable vacillation, since its selective acts are never definitively decided; ... the aesthetic object is always, vis-à-vis the selection of signifiers, both signifier and material" (p. 44). So aesthetic experience is a "self-subversion" or "processual deferral" of signifier formation.

This aesthetic deferral displays three features, according to Menke. First, signifiers already automatically selected get reduced to attributes of the materials, leading to a "supplementation of unselected material" (p. 47). Second, there is a disruption to contexts that usually provide criteria for settling nonaesthetic disruptions of meaning

---

[8] Although not obvious in the English translation, "the negated" in this sentence refers to understanding itself, not to the artwork that one is attempting to understand. See Menke, *Die Souveränität der Kunst* (1991), p. 44.

(e.g., when a sentence is grammatically flawed). Third, aesthetic experience *quotes* rather than *applies* contextual assumptions, with a double result: contextual assumptions become ambiguous, and signifiers acquire "an unsublatable indeterminacy" (p. 60). Hence it is "precisely *in* the attempt at aesthetic signifier formation that the aesthetic object produces itself as material and achieves its superabundant quality vis-à-vis each and every signifying selection" (p. 61). This superabundance is a "surplus" (Derrida), a "superabundance of meaning" accruing to signifiers in, and only in, a process of aesthetic deferral.

In chapter 4 Menke explores the implications of aesthetic negativity for interpretation and evaluation. Aesthetic interpretation has the task of expressing aesthetic experience, he claims. To do this, interpretive speech must evaluate "whether an aesthetic experience is stringent in its negativity and whether the statements of an interpretive speech are discontinuous" (p. 126). Similarly, aesthetic evaluation does not apply substantive criteria to the aesthetic object but rather assesses the stringency of the aesthetic experience itself:[9] "An object that is stringently experienced (or can be so experienced) in an aesthetical manner is aesthetically good; an object that is not stringently experienced (or cannot be so experienced) in an aesthetical manner is aesthetically poor" (p. 131). Hence, aesthetic experience "is a process of direct normative experience" (p. 140).

Postaesthetic Subversion

Part II (chapters 5–8) takes up the topic of aesthetic sovereignty as postaesthetic subversion. Menke acknowledges that his account of autonomy in part I runs the risk of isolating art. From a Derridean perspective, limiting the power of aesthetic negativity "to the narrow province of aesthetic experience" (p. 161) would pose a problem. To limit and neutralize negativity would conflict with Derrida's claim that art is "sovereign" in its ability to threaten meaning-producing discourses. For Derrida, "[a]rt becomes sovereign if the experience of

---

[9] Menke claims that Adorno's attempts to ground aesthetic evaluation in objective technical and historical criteria tend to undermine the autonomy of aesthetic experience by subjecting stringency to nonaesthetic norms.

its negativity at the same time uncovers the hidden negativity *also* found ... in functioning [nonaesthetic] discourse" (p. 165).[10]

In response to this problem, Menke introduces a crucial distinction between "implications" and "consequences": whereas Derrida takes aesthetic negativity to *imply* a deconstruction of nonaesthetic discourses, Adorno regards aesthetic negativity as *having consequences* for one's picture of nonaesthetic discourses. Menke prefers Adorno's approach because it retains the autonomy of aesthetic experience. Yet Adorno does not ignore art's sovereignty, Menke claims, insofar as Adorno regards aesthetic experience "as being *potentially ubiquitous*" (p. 172). This viewpoint, in turn, requires the "affirmation of an aesthetic experience no longer institutionally situated." So the central difference concerning aesthetic sovereignty comes to this: whereas Derrida thinks that sovereign aesthetic experience implies "nonaesthetically valid insight into the negativity of all discourses," Adorno thinks that "sovereign aesthetic experience produces a postaesthetic subversion of our discourses" (pp. 178–9).

Menke derives the notion of postaesthetic subversion from the critique of enlightenment rationality in Adorno's *Negative Dialectics*. Like Derrida, Adorno thinks reason is caught by an irresolvable and negative dialectic "between reason's infinite claims and its finite means" (p. 211), as seen, for example, in the Hegelian claim to absolute knowledge of absolute truth. Adorno goes beyond Derrida, however, in demonstrating why reason *must* raise infinite claims. The grounds reside in an emphatic concept of experience, which, according to Menke, Adorno interprets in a genealogical way: when Adorno grounds the negative dialectic, he focuses "not on the satisfaction of infinite claims, but on their genesis" in "a fundamental experience of a crisis of reason" (p. 216).

Menke takes up two versions of this generative experience. The first is the modern experience of death as something that metaphysics can no longer claim to comprehend (p. 218). Adorno thinks this experience thoroughly disrupts rational discourses, forcing them to raise

---

[10] Menke suggests that, by finding traces of aesthetic negativity in nonaesthetic discourses, Derrida transforms both aesthetic signs and nonaesthetic discourses into texts: deconstructive terms such as "différance" and "dissémination" are "structures of aesthetic experience with generalized validity" (p. 167).

infinite claims for themselves that cannot be satisfied. Against Adorno, Menke argues that *at most* the modern experience of death can render discursive achievements *irrelevant* to participants in discourse. It does not fundamentally challenge the validity of nonaesthetic discourses.

That leaves the second version of the generative experience, one that Menke thinks really does plunge "our discourses and practices into a crisis *from the outside*, but nonetheless a crisis *for them*" (p. 223). The generative experience in question is the aesthetic experience of negativity. On this reading of Adorno, reason's infinite claims "arise out of the confrontation of our discourses with the aesthetic experience of negativity" (p. 226). Unlike nonaesthetic challenges to reason, aesthetic experience can be a *total* negation of our discourses to the extent that "the aesthetic attitude becomes a general stance" (p. 230). As Menke explains, the aesthetic attitude transfigures discursive representations, estranging them from their nonaesthetic contexts. When this shift becomes a general stance, it subverts the validity of nonaesthetic discourses. Nothing can limit the scope of such subversion. It affects not only the relevance of nonaesthetic discourses but also their successful functioning. In response, discursive rationality must raise "absolute claims to meaning and grounding" (p. 237), claims that rational discourses can never satisfy: "Only the potential ubiquity of the aesthetic experience of negativity contorts the calm face with which rational discourses turn toward their immanent analysis into the (negative-)dialectical grimace of a reason that is forced to make excessive demands of itself to repulse aesthetic negativity" (p. 239).

Accordingly, where Adorno claims that, because of their "functionlessness," modern artworks challenge a "bewitched reality" and negatively embody utopia (*AT* 227/336–7), Menke says that they give rise to negative aesthetic experience whose ubiquity challenges contemporary rationality. Whereas Adorno ties art's social function to hope for societal transformation, Menke ties this to the crisis of rationality's selftransgression. Whether Menke's aesthetics of negativity is a viable reconstruction remains to be seen.

1.2 LIBERATION AND DECONSTRUCTION

To set a stage for questioning Menke's reconstruction of Adorno, I need to make two comments on the multivalence of "autonomy" as a central concept in Western aesthetics since the eighteenth century. First, as Casey Haskins points out, autonomy has served both justificatory and explanatory roles. In its justificatory employment, autonomy grounds arguments for "the uniqueness of art, or aesthetic experience, or both, as a *source of value*, and their logical irreducibility to other sources of value."[11] Put to explanatory use, by comparison, autonomy grounds arguments to the effect that art or aesthetic experience cannot be properly explained in terms of economic, political, psychological, or other factors. Common to both uses is the notion that art or aesthetic experience is irreducible, usually on formal grounds – by virtue of following its own rules or principles or "logic." In the late nineteenth century this notion of irreducibility links up with claims for art's superiority to other modes of discourse such as science, religion, and bourgeois morality (e.g., in the "art for art's sake" movement). This is the historical site for the intertwining of autonomy with sovereignty that Menke has unraveled and rewoven. In the twentieth century claims for art's superiority or sovereignty join modernist emphases on the self-referential purity of each artistic medium, whether poetry, painting, music, or modern dance (e.g., New Criticism in literature, Clement Greenberg's Modernist Criticism in the visual arts, and much of the New Music movement).

Second, the case for autonomy, whether justificatory or explanatory, has varying points of reference. In general, the case can apply to either the aesthetic dimension or the arts. Those who argue for the autonomy of the aesthetic dimension usually do so via an account of aesthetic experience or aesthetic judgment (e.g., Immanuel Kant), although it is possible to ground autonomy in the objects of aesthetic experience instead (e.g., Mikel Dufrenne, on a certain reading). Those who argue for autonomy with respect to the arts can do so in terms of the work of art (e.g., Stéphane Mallarmé), the artist (e.g., through an appeal to "genius"), the experiencing of art by audiences, viewers, readers, and the like (e.g., certain reader

---

[11] Casey Haskins, "Autonomy: Historical Overview," in *Encyclopedia of Aesthetics*, ed. Michael Kelly (New York: Oxford University Press, 1998), 1:170–5; quotation from p. 170.

response theories), or art itself as an entire branch of culture (e.g., G. W. F. Hegel).[12]

From here on I shall distinguish two intersecting lines drawn from these points of reference. I shall use the term "aesthetic autonomy" for a concept that locates autonomy (somewhere) in the aesthetic dimension, and the term "artistic autonomy" for a concept that locates autonomy in (some aspect of) the arts. Kant's *Critique of Judgment* provides somewhat independent accounts of the aesthetic and the artistic and, by extension, of their autonomy. Such independence becomes the exception rather than the rule, however, once Hegel declares artistic beauty far superior to the beauty of nature and equates "aesthetics" with "the philosophy of fine art."[13]

Double Reduction

Menke's reconstruction of Adorno reflects this Hegelian background. Despite devoting a chapter to the concept of beauty, Menke does not acknowledge any aesthetic experience other than that which is tied to the experience of art. Moreover, because he rejects Adorno's "objective" criteria for evaluating artworks, aesthetic experience becomes a free-floating experience whose autonomy rests in its "negativity" toward other discourses and whose sovereignty rests in its "potential ubiquity."

These moves involve a double reduction that is doubly problematic. Menke reduces artistic autonomy to aesthetic autonomy, and he reduces aesthetic autonomy to the supposed negativity of aesthetic experience. I find this double reduction problematic both because it vitiates the emancipatory potential of Adorno's aesthetics and because it turns artistic autonomy into a principle of illusory transgression. Let me first demonstrate the double reduction before I show in more detail why it is doubly problematic.

---

[12] The first and last of these – the artwork and art as a branch of culture – play an especially important role in the history of aesthetics. See Peter Bürger, "Critique of Autonomy," in *Encyclopedia of Aesthetics*, 1:175–8.

[13] An important innovation in Adorno's *Aesthetic Theory* is his resuscitating the topic of natural beauty without abandoning a modernist focus on art as such. Cf. Heinz Paetzold, "Adorno's Notion of Natural Beauty: A Reconsideration," in Huhn and Zuidervaart, *The Semblance of Subjectivity*, pp. 213–35.

The first reduction – the reduction of artistic autonomy to aesthetic autonomy – occurs in part II of Menke's book. There he argues that art is sovereign by virtue of its being constituted in a modern, differentiated society as the one discourse where all rational discourses, including itself, are thrown into permanent crisis. The challenge that art poses to understanding occurs in the very process of understanding. Upon closer examination, this account of art's sovereignty concerns not so much the arts themselves as the experience people have when they enjoy, interpret, and evaluate the arts. Further, Menke describes only one portion of such experience, namely, what he calls "aesthetic experience." If, however, one subscribes to the view that the arts are multidimensional, as are the modes in which they are experienced, then Menke's account of art's sovereignty appears both monological and reductive: monological, because it restricts the structure of art to its aesthetic dimension; and reductive, because it limits the experiencing of art to an aesthetic experience. In order to credit modern art with far-reaching deconstructive potential, Menke must set aside the ways in which the arts participate in political, economic, moral, religious, and even scientific processes, and he must ignore the ways in which the experiencing of art normally and properly includes nonaesthetic concerns.

The second reduction – that of aesthetic autonomy to the supposed negativity of aesthetic experience – occurs in part I of Menke's book, where he describes aesthetic experience as experience in which pleasure arises from the unavoidable subversion of equally unavoidable attempts to understand an aesthetic object. His description implies that neither the object nor any context of understanding nor any substantial standard governs the "logic" of aesthetic experience. Rather, the autonomy of the aesthetic dimension depends *entirely* on whether the aesthetic process of negation is "stringent." Moreover, the only criterion for stringency seems to be whether an aesthetic experience is thoroughly and consistently disruptive of "automatic understanding." This theory of aesthetic autonomy must discount traditions of aesthetic interpretation, institutionalizations of aesthetic conduct, and norms of intersubjective aesthetic discourse. But to theories that credit these as sources of positive content to aesthetic experience, Menke's account of aesthetic autonomy appears both derivative and reductive: derivative, because it derives the normativity of aesthetic experience from a subversion of established and primarily nonaesthetic norms; reductive, because it reduces the content of aesthetic experience to a process of interminable negation. In order to anchor autonomy in the negativity of aesthetic experience, Menke must ignore the degree to which aesthetic experience is itself constituted by its sociohistorical structuration. He must also understate the internal complexity of the aesthetic dimension.

Emancipatory Potential

As I have already suggested, Menke's double reduction is problematic as a reading of Adorno. Menke himself acknowledges that his reconstruction renders inoperative several of Adorno's central claims. Let me focus on two. The first concerns the utopian reach of modern art; the second, the notion of aesthetic understanding.

When Menke rejects nonaesthetic grounds for a negative dialectic, he sidesteps altogether the role of suffering in Adorno's critique of enlightenment rationality, a topic I take up at greater length in the next chapter. Not only does Adorno regard the need to express suffering "a condition of all truth" (*ND* 29/17–18) but he also considers such expression central to the import of modern art (*AT* 260–1/386–7). Recognition of suffering and hope for relief motivate much of emancipatory thought, from Hegel's dialectic of master and slave through Karl Marx's critique of capitalism to Georg Lukács's diagnosis of reification and Adorno's negative dialectic. None of these critical projects make much sense apart from a utopian projection, even if only implicit, of a social condition in which oppression has ended and suffering has been relieved.

Accordingly, I think it is fundamentally mistaken to interpret Adorno as attributing art's deconstructive potential simply to its capacity for aesthetic negativity. To the extent that art in its modern differentiated condition throws rational discourses into "permanent crisis," it does so by expressing unmet needs and unfulfilled desires in ways that not only recall the limits of rationality's achievements but also suggest both the hope and the possibility of surpassing these limits.

I realize, of course, that saying this in Adorno's defense lays one open to Jürgen Habermas's critique of aestheticism, which I discuss in Chapter 4. In *The Philosophical Discourse of Modernity* Habermas accuses Adorno of structurally overburdening the achievements of art in modernity: because Adorno expects art to solve problems of rationalization that no differentiated value sphere, whether art, science, or morality, can solve single-handedly, he overlooks the potential for communicative rationality released by modern rationalization. Instead, as Menke summarizes Habermas's critique, Adorno postulates a "remythicizing" solution to reason's problems in a "transrational" aesthetic experience (p. 247). Paradoxically, rationalization itself makes possible this supposedly salvific aesthetic experience.

Yet there is a problem with Habermas's critique. As Menke points out in his final chapter, Habermas mistakenly reads Adorno as having a romantic rather than a modern conception of art. Whereas the romantic conception (e.g., Schelling) positions art teleologically beyond modern rationality as a solution to its aporias, Adorno – at least in those formulations Menke embraces – positions art genealogically as "the catalyst of problems that cannot even arise or be conceived of without aesthetic experience" (p. 249). Art is not a problem solver for reason's crisis; it is the source of that crisis.[14]

According to Menke, then, Adorno does not remove art from its differentiated status. Yet Adorno does challenge Habermas's proposal to calm reason's discontents via an integrating interplay among scientific, moral, and aesthetic rationality. Habermas's proposal would avoid art's sovereignty altogether by neutralizing art's negativity. Art's negativity, Menke claims, "is in no relationship of interplay with nonaesthetic reason but ... in a relationship of interminable crisis" (p. 254).

But precisely here Menke's reading of Adorno flounders. An interminable crisis is not the same thing as a significant change. While I agree with Menke that surpassing rationality's achievements

---

[14] In this context Menke makes the illuminating suggestion that Derrida's account of aesthetic sovereignty is a form of "reverse romanticism": it removes aesthetic experience from modern differentiation and elevates it above nonaesthetic discourses, yet "no longer characterizes aesthetic experience as the medium of reconciliation, but as that of interminable deferral" (pp. 250–1).

does not require irrational transcendence, I believe Adorno sees this occurring through a transformation of rationality itself. Transformation is not to be equated with mere subversion, nor is the negativity of modern art *mere* negativity. What throws modern art into crisis, according to Adorno, and what accounts for its critical capacities toward discursive rationality, is an unavoidable conflict between art's internal freedom and the perennial unfreedom of modern society as a whole (*AT* 1–2/9–11).

My second criticism of Menke's interpretation concerns the notion of aesthetic understanding. When Menke describes aesthetic negativity as understanding's unceasing self-subversion, he discards two central notions in Adorno's account of aesthetic understanding, namely, truth content (*Wahrheitsgehalt*) and technique. Adorno does not valorize the pleasure supposedly derived from disrupted understanding. Nor does he anchor autonomy in aesthetic experience. Rather, he thinks that aesthetic experience *achieves* autonomy to the extent that it does justice to the autonomy of the *object* of aesthetic experience, whether that be natural beauty as a socially mediated trace of the nonidentical or the artwork as a peculiar product of human labor. Even his highly self-reflective essay "Trying to Understand *Endgame*" assumes that the unintelligibility of Samuel Beckett's play can be understood: Adorno seeks the object of understanding in the play's truth content, as constituted and mediated by its technique.[15] This implies that the objects of aesthetic understanding, or at least those artworks which are such objects, provide their own basis for interpretation. In addition, their evaluation can appeal to criteria that pertain to the *objects* themselves and not simply to the quality ("stringency") of the *aesthetic experience* they occasion.[16]

Adorno's locating of autonomy in the object is crucial, because it allows him to argue that artworks can contain and foster social criticism. He insists on the artwork's autonomy in order to maintain a

---

[15] See Theodor W. Adorno, "Trying to Understand *Endgame*," in *Notes to Literature*, vol. 1, ed. Rolf Tiedemann, trans. Shierry Weber Nicholsen (New York: Columbia University Press, 1991), pp. 241–75; "Versuch, das Endspiel zu verstehen," in *Gesammelte Schriften*, vol. 11 (Frankfurt am Main: Suhrkamp, 1974), pp. 281–321. I comment on this essay in "Paradoxical Modernism," in *Adorno's Aesthetic Theory*, pp. 150–77.

[16] This is an aesthetic version of Adorno's general insistence on the "priority of the object," which I discuss in Chapter 2.

space for social critique in a society where oppositional forces are regularly co-opted, ignored, or marginalized. Menke, however, sets aside "social-critical" accounts of aesthetic negativity (pp. 7–12), strips the "metaphysical" layers from Adorno's conception of aesthetic meaning (pp. 89–91), and surrenders Adorno's technical and historico-philosophical criteria for evaluating artworks (pp. 132–6). He thereby loses sight of the reasons why Adorno would consider autonomy to be so crucial, namely, as a way to preserve social-critical and utopian capacities in a late capitalist society. Menke's double reduction vitiates the emancipatory potential of Adorno's aesthetics.

Illusory Transgression

Because I have reasons to question Adorno's conception of autonomy, however, my problematizing of Menke's reading does not suffice as a criticism of his double reduction. Countering my defense of Adorno, one could say that the emancipatory thrust of his aesthetics is itself problematic. Hence, weakening it in a Menkian manner is not too high a price to pay.

Given that possible objection, let me propose a second and more systematic line of criticism. I wish to argue now that Menke's double reduction turns artistic autonomy into a principle of illusory transgression. To develop this argument, I shall reverse the order followed in my defense of Adorno, beginning this time with the reduction of aesthetic autonomy to aesthetic negativity in part I of Menke's book and then proceeding to the reduction of artistic autonomy to aesthetic autonomy in part II.

Conventions and Content
When I demonstrated Menke's reduction of autonomy to negativity, I said that he derives the normativity of aesthetic experience from the subversion of nonaesthetic norms and hereby reduces the content of aesthetic experience to a process of interminable negation. I simply assumed that aesthetic experience can have its own norms and content. My assumption seems incompatible with Menke's version of the semiotic turn in aesthetic theory. That leaves at least two options: either to reject the semiotic turn altogether, or to argue

that a semiotic account need not reduce aesthetic autonomy to aesthetic negativity. I shall pursue the second option.

Menke's semiotic account locates the fundamental difference between aesthetic and nonaesthetic experience in the claim that aesthetic experience does not rely on conventions for identifying objects and does not establish a meaningful link between the material and the sign. In a way, he has simply updated Kant's claim that judgments of taste do not apply definite concepts to objects and do not rely on a conceptually mediated interest in an object's existence.[17] Just as I have never found Kant's claim persuasive, so I have difficulty accepting Menke's way of differentiating between aesthetic and nonaesthetic experience.

Let me limit my remarks to the matter of conventions and positive content in aesthetic experience. Take as an example the practice in Western societies of going for a walk to enjoy the sights and sounds and smells of the outdoors. This is the sort of activity in which aesthetic experience and Kantian judgments of taste commonly occur, and it cannot simply be equated with experiencing artworks as aesthetic objects. At some point while walking, or perhaps throughout the walk, people will experience the setting or occasion as "beautiful" or "sublime" or "picturesque," to use an eighteenth-century vocabulary. Often one will turn to one's companions, or perhaps to oneself, and remark on how wonderful the scene or the occasion is. At that point one has, in effect, tried to reach an understanding of what is being experienced as an aesthetic object or event.

Why must such an experience be characterized as primarily nonconventional and negative? Perhaps the practice of "going for a walk" is framed negatively in industrialized and increasingly computerized societies – as an escape from routine or a way to relieve stress, for example – but then the interesting point would be that the practice *is framed*. It is a conventional practice with its own social history and internal rules. In fact, cities and nations set up special spaces where the practice has better chances of succeeding.

---

[17] Immanuel Kant, *Critique of the Power of Judgment*, ed. Paul Guyer, trans. Paul Guyer and Eric Matthews (Cambridge: Cambridge University Press, 2000), §§1–5, pp. 89–96.

What could it mean, then, to say of this situation, using Menke's terms, that the signifier interminably vacillates between the poles of material and meaning? I see little vacillation. For purposes of semiotic analysis, we can distinguish at least three elements in the situation described: the humanly perceived landscape or some aspect of it, the vocalization of the sounds "That's gorgeous," and the speech act of saying to someone or to oneself that this landscape is gorgeous. Although it would be misguided to say the speech act is primarily assertoric,[18] it would also be phenomenologically inaccurate to say that the speech act has no assertoric dimension whatsoever. "Gorgeousness" may not be a predicate for which we have ready-made rules of application, but it is not altogether lacking in content nor completely unbound from conventions of usage. The aesthetic object (the humanly perceived landscape) is such that one vocalization is more appropriate than another in a given language, and the vocalization is such that it is not hard to establish a meaningful link between material and sign.

Perhaps Menke wishes to suggest that aesthetic experience is fundamentally nondiscursive, and that the linguistic enunciation of interpretations and understandings is not intrinsic to aesthetic experience. But I find that suggestion dubious, and it raises questions as to why Menke himself wishes to give aesthetic theory a semiotic turn. The only situations I can imagine where Menke's account would sometimes hold good are when people try to understand certain works of modern or contemporary art. Yet even there it is implausible to claim that "aesthetic signs are not used in any context" (p. 53). At a minimum, people use them in the conceptual and historical context provided by what Arthur Danto calls "the artworld."[19]

---

[18] To that extent, but only to that extent, Kant is correct when he describes taste judgments as noncognitive.
[19] In addition to the books by Danto listed in the Bibliography, one should consult Noël Carroll's lucid essay "Essence, Expression, and History: Arthur Danto's Philosophy of Art," in *Danto and His Critics*, ed. Mark Collins (Oxford: Blackwell, 1993), pp. 79–106, and the brilliant critical introduction by Gregg Horowitz and Tom Huhn in their compilation of essays by Danto titled *The Wake of Art: Criticism, Philosophy, and the Ends of Taste* (Amsterdam: G + B Arts International, 1998), pp. 1–56.

Contrary to Menke's account, my example suggests that aesthetic experience of the most ordinary sort has positive content and its own conventions. It does not need to subvert nonaesthetic norms or repeatedly disrupt "automatic understanding" in order to have autonomy. What gives aesthetic experience a certain degree of autonomy is the fact that in modern differentiated societies an entire network of practices has developed to provide settings and vocabularies and shared activities within which and through which people can (learn to) play and decorate and fashion styles of life with aesthetic aims in view.

Moreover, these aims do not pertain in the first instance to a negation of understanding but to the enactment of intersubjective processes of imagination. By "imagination," as is explained later, I mean all the ways in which people can explore, present, and creatively interpret aesthetic signs. As "aesthetic signs," many different creatures, events, and products can "make multiple nuances of meaning available in ways that either exceed or precede both idiosyncratic expressions of intent and conventional communications of content." Although aesthetic signs are misinterpreted if they are read as having logical reference, they do call for explications that presuppose them to be "about something other than themselves." Such "aboutness" is "both sharable and shared by various interpreters."[20] As imaginative "referral," the aboutness of aesthetic signs cannot be captured in Derridean "deferral," no more than, as the intersubjective hermeneutic of aesthetic signs, imagination can be reduced to the idiosyncratic and interminable disruption of "automatic understanding." In fact, it makes little sense to claim, à la Derrida, that in aesthetic deferral a "surplus" or "superabundance of meaning" accrues to signifiers. If, according to Menke, this surplus arises from a reduction of signifiers to materials and from a disruption of the contexts needed to achieve meaning, how can the "surplus" be a superabundance *of meaning*? Nor is it clear why one should describe aesthetic objects as "aesthetic signifiers" if no meaningful link can be established with what they signify. Is a signifier definitively decoupled from the signified still a signifier?

---

[20] Lambert Zuidervaart, *Artistic Truth: Aesthetics, Discourse, and Imaginative Disclosure* (Cambridge: Cambridge University Press, 2004), pp. 60–1.

Hence I question Menke's claim that aesthetic pleasure arises from the way in which aesthetic experience transforms nonaesthetic acts of recognition and releases an "unsublatable negativity" (pp. 13, 25). In ordinary aesthetic experience, our wonder or puzzlement or shock arises just as much from the release of an "unsublatable" positivity – the flip side, if you will, to Adorno's notion of "the nonidentical." While I agree that aesthetic pleasure arises in "reflective recourse ... to the process of experiencing the object" (p. 13), this process is not one of perpetually deferred understanding. Rather it is one in which understanding often is both relatively direct and somewhat noncontrovertible. Such understanding may disrupt other conventions and contexts of signification and thereby acquire negativity. But this negativity is not as interminable as Menke suggests, for aesthetic understanding itself relies on conventions and contexts provided by a network of aesthetic practices in modern differentiated societies. Although the network's pervasiveness helps explain the "potential ubiquity" of an aesthetic stance, this need not create an "interminable crisis" for rationality if, in fact, aesthetic experience is not reducible to aesthetic negativity.[21]

Technique and Technologies
Turning now to Menke's reduction of artistic autonomy to aesthetic autonomy, let me recall my earlier claim that Menke restricts the structure of art to its aesthetic dimension and limits the experience of art to an aesthetic experience. A fundamental problem in Menke's account is his presupposing without sufficient argument that, to be both "autonomous" and "sovereign," art must also be one-dimensional. To regard art as primarily or exclusively aesthetic, one must ignore the obvious fact that art making and art

---

[21] Not only does Menke exaggerate the nonconventional and negative character of aesthetic experience, but he also ignores the nonconventionality and negativity of nonaesthetic discourses. Although I agree with many of Menke's criticisms of Derrida, I think Derrida is more attuned than he to the nonconventional and negative within nonaesthetic discourses, as are Julia Kristeva and Adorno, for that matter. Adorno never restricts the "mimetic" dimension of discourse to the arts and aesthetic experience. In this connection, see Martin Jay, "Mimesis and Mimetology: Adorno and Lacoue-Labarthe," and Shierry Weber Nicholsen, "*Aesthetic Theory*'s Mimesis of Walter Benjamin," in Huhn and Zuidervaart, *The Semblance of Subjectivity*, pp. 29–53 and 55–91, respectively.

experiencing unavoidably depend on media and on methods and means of production and distribution.

For this reason, an account of art that affirms "an aesthetic experience no longer institutionally situated" (p. 177) is highly implausible. Art cannot subvert discursive rationality unless it also challenges the prevailing modes of production and distribution and does so within and through art's own technical dimension. I take this to be a central insight in Walter Benjamin's overly optimistic account when, for example, he writes: "The technical reproducibility of the work of art changes the stance of the masses toward art. The most reactionary stance, for example, toward a Picasso painting, is replaced by the most progressive, for example, toward a Chaplin movie."[22] Adorno reworks this insight into the less optimistic but nonetheless utopian claim that "[p]raxis is not the effect of [art]works; rather it is encapsulated in their truth content" (*AT* 247/367).

To elaborate this point, let me introduce a distinction between technology and technique. I use "technology" to refer to those instruments and procedures of production and distribution which the arts share with other fields of human endeavor in industrial and postindustrial societies. Film and photographic reproduction were clear examples in Benjamin's day, as are televisual and computer technologies today. I use "technique" to refer to those media and methods which are somewhat unique to specific forms of art. The methods of twelve-tone composition in music and montage in the visual arts provided prominent examples when Adorno was developing his philosophy of the arts; fragmentary repetition in minimalist music (e.g., Philip Glass) and site-specific installations in the visual arts (e.g., Christo) provide more recent examples. Social technology and artistic technique interact and intersect. Together they constitute the technical dimension of the arts.[23]

---

[22] Walter Benjamin, "The Work of Art in the Age of Mechanical Reproduction," in *Illuminations*, ed. Hannah Arendt, trans. Harry Zohn (New York: Schocken Books, 1969), pp. 217–51; quotation from p. 234. "Das Kunstwerk im Zeitalter seiner technischen Reproduzierbarkeit," in *Illuminationen: Ausgewählte Schriften* (Frankfurt am Main: Suhrkamp, 1977), pp. 136–69; quotation from p. 159. Translation modified.

[23] Whereas Benjamin emphasizes the progressive impact of new social *technologies* on art production and reception and thereby on people's political consciousness,

No contemporary philosophy of the arts can afford to ignore the constitutive role of the technical dimension in the formation of artistic autonomy. A justificatory argument concerning artistic autonomy must show that in some sense the arts are unique and irreducible sources of value *in their technical dimension*. Similarly, in an explanatory vein, one needs to argue that *the technical dimension* of the arts cannot be properly explained in terms of other social factors. The strength of Adorno's account lies precisely here. Adorno justifies artistic autonomy, in part, by arguing that authentic modern technique in the arts reverses the direction of instrumental rationality in capitalist societies. Rather than imposing universal solutions on neutralized "materials," modern artists find specific problems in their historical media and elicit imaginative solutions, without covering up the tensions that remain. Further, when Adorno explains artistic autonomy, he insists on the uniqueness of artistic technique, without denying its connections with social technology. For these reasons, as I have shown elsewhere, Adorno thinks that modern art making can model how all labor could be transformed, from a condition of alienation from nature and of economic exploitation to one of disalienation and social solidarity.[24]

In the decades since Adorno's death, however, it has become more apparent how large a role new social technologies play in the development of artistic media and methods. One thinks, for example, of the impact of computer technologies on the composition, performance, recording, dissemination, and reception of music. Many commonplace sounds and "manipulations" of sound in contemporary music would be unthinkable in the absence of computer technologies. We no longer live in what Benjamin describes as an age of technical reproducibility, in which discrete *works* of art can be reproduced as photographic images or sound recordings. Today the very *practice* of art making is endlessly reproducible in thousands of permutations. In a sense, any computer user can be a composer or poet or digital artist for a vast potential audience. Sometimes computers themselves can do most of the work. Technologically,

Adorno stresses the role of new artistic *techniques* in constituting artworks that have a progressive import.
[24] See Zuidervaart, *Adorno's Aesthetic Theory*, pp. 109–21.

then, as well as economically, art making has entered an age of global reproducibility. This raises significant questions about the autonomy of art, in both justificatory and explanatory respects.

Be that as it may, it should be clear why I reject Menke's reduction of artistic autonomy to aesthetic autonomy. The reason is quite simple. The arts are not simply fields of aesthetic activity and experience. They are also, at a minimum, and necessarily, fields of technical activity and experience. To be irreducible to other sources of value or to other social factors, they will need to be irreducible in their technical dimension. Whether or not a plausible case can be made for technical irreducibility, Menke's reduction of artistic to aesthetic autonomy proves untenable. If no plausible case can be made, this would suggest that the arts are not autonomous, not even in their aesthetic dimension. If, however, a plausible case can be made for the technical irreducibility of the arts, then this would indicate that the arts and their autonomy are at least two-dimensional, and the reduction of artistic autonomy to aesthetic autonomy is inappropriately one-dimensional.

When one combines my systematic criticisms of Menke's two reductions, art's purported "autonomy" and "sovereignty" turn out to be much less deconstructive than Menke claims. In the first place, aesthetic experience is more conventional and positive than he suggests. In the second place, art is more complex and socially embedded than he acknowledges. If contemporary arts in their aesthetic dimension are to challenge other rational discourses, they will need to disturb the conventions and content of ordinary aesthetic experience through media and methods of art making that are not merely aesthetic. Otherwise their challenges will have little resonance among contemporary publics and will not address the societal formation that nonetheless makes them possible. They will be nonnature preserves of a free-floating negativity. Whatever disturbance they create will occur at a relatively safe distance. Their autonomy will be a principle of illusory transgression, as useful to savvy marketers and media moguls as it is irrelevant to a process of societal transformation.

I do not believe that either Adorno's aesthetic modernism or Menke's postmodern aestheticism provides an adequate account of autonomy in the age of art's global reproducibility. Neither one has a satisfactory view of how artistic practices and organizations can both contribute to the transformation of contemporary society and, in contributing, retain and strengthen their intrinsic worth. This strikes me as a central conundrum, both theoretical and practical, for an aesthetics after Adorno. As such, it gives debates about artistic "autonomy" or "sovereignty" their social-philosophical point. By way of a conclusion, and as a transition to subsequent chapters, let me sketch an alternative view.

1.3 AESTHETIC AND ARTISTIC AUTONOMY

Central to my criticisms of Menke is a concern to distinguish the aesthetic from the artistic, and to show how they are related. My concern stems in part from Adorno's criticisms of Hegel for collapsing that distinction.[25] Adorno returns to Kant in this regard, but not in the way that Menke does. For Adorno takes seriously the Kantian concept of "natural beauty" (*die Naturschönheit*) and does not subordinate it to Hegel's "artistic beauty" (*das Kunstschöne*). While it is so that Adorno has a negative dialectical approach to both categories, and in that sense offers an "aesthetics of negativity," the salient point for present purposes is that he retains the distinction, and he considers it important both for art and for a society that would not be "bewitched."

Accordingly, when I previously discussed the example of going for a walk, I did so in order to make room for "aesthetic experience" that is not directly arts related. For the many philosophers who do not accept my distinction between the artistic and the aesthetic, this example could raise a number of questions. At least three come to mind. (1) Is there significant continuity between aesthetic experience of "nature" and aesthetic experience that is arts related? (2) Is not such aesthetic experience (of "nature"), although conventional, nevertheless fundamentally negative? (3) Would not my analysis of

---

[25] The concern also arises from the less-well-known work of my mentor Calvin Seerveld, who developed a "modal aesthetics" to account for the aesthetic dimension of life, culture, and society. Seerveld argues that, although prominent in the arts, the aesthetic dimension is also found throughout nonartistic experiences, practices, and institutions. See, for example, Calvin Seerveld, "Imaginativity," *Faith and Philosophy* 4 (January 1987): 43–58.

this example, if transferred to the aesthetic experience of artworks, implicitly endorse that sort of naive "art enjoyment" which Adorno, and most modern and postmodern artists too, would find both artistically and politically problematic? My brief response, which clearly needs elaboration, is as follows. (1) There is sufficient continuity to justify Adorno's claim that "reflection on natural beauty is irrevocably requisite to the theory of art" (*AT* 62/98). (2) The "negativity" of garden-variety aesthetic experience is itself bound up with an "image" of that which is not negative – "nature as the mediated plenipotentiary of immediacy," in Adorno's terms (*AT* 62/98). (3) My incomplete analysis of this example cannot simply be transferred to arts-related aesthetic experience, because artworks have aesthetically relevant technical aspects in ways that landscapes do not. This is not to deny that landscapes where people go for walks are, to a significant degree, "cultural" landscapes and not simply "natural," but to recognize that most landscapes of that sort are not art*works*, the earthworks of Robert Smithson and Andy Goldsworthy notwithstanding.

Imagination

Beyond positing a distinction, however, one needs to give a sufficiently comprehensive characterization of aesthetic matters both to include that which is not artistic and to show why the aesthetic is crucial to art's distinctive tasks in a differentiated society. Because I have already attempted this in my book *Artistic Truth*, let me simply summarize some of the arguments given there and show how these accomplish a critical retrieval of Adorno's insights. In a chapter titled "Kant Revisited" I offer a partial phenomenology of the aesthetic dimension, in order to understand how artworks and art talk can help people find cultural orientation. Reading Kant's aesthetics as a dialectical road map to an emerging institutionalization of aesthetic matters in modern Western societies, I characterize "the aesthetic" in terms of three intersubjective processes: exploration, whose concept surfaces in a dialectic between work and play; presentation (*Darstellung*), whose concept arises from a dialectic between expression and communication; and creative interpretation, whose concept stems from a dialectic between entertainment and instruction. Taken together, exploration, presentation, and creative interpretation are central to what has become differentiated and institutionalized as an aesthetic dimension. I consider the differentiation of this dimension a historical achievement whose reversal, subversion, or transfiguration would require that an entire societal formation be transformed.

My general term for these three processes, and for the aesthetic dimension as a whole, is "imagination." I do not understand this as a mental capacity à la Kant (*Einbildungskraft*), but rather as a complex of intersubjective processes tied to a wide and changing array of institutions and practices, such as "going for a walk to enjoy the sights and sounds and smells of the outdoors." On my account, an experience like this is inherently intersubjective and unavoidably hermeneutical. It involves elements of exploration, presentation, and creative interpretation. Here I share Menke's intuition, and Adorno's too, that the enjoyment people take in aesthetic experience is "reflective" rather than immediate. Yet I tie such reflexivity to the intersubjective and hermeneutical character of the experience, rather than to its capacity, which I do not completely deny, to transform "automatic acts of recognition." For I regard most *non*aesthetic understanding to be intersubjective and hermeneutical as well, although not primarily imaginative, and hence to be less "automatic" than Menke suggests.

Keeping in mind Adorno's insistence on the "priority of the object," I also acknowledge that certain entities are more suited than others, and certain features of entities are more suited than other features, to enter processes of exploration, presentation, and creative interpretation. When they enter imaginative processes, they do so as aesthetic signs. I do not claim that the "objects" of imagination *are* aesthetic signs. Rather, I say that various entities *can function as* aesthetic signs when they enter certain intersubjective processes. The term "aesthetic signs" refers to "creatures, events, and products" in their capacities to "sustain discovery," to "acquire nuances of meaning," and to "call forth reflective readings," all of this occurring "in intersubjective contexts."[26]

---

[26] Zuidervaart, *Artistic Truth*, pp. 61–2.

On my account, the processes of imagination are not anormative or antinormative. Nor does their normative character derive from the "stringency" with which they disrupt "automatic understanding," as Menke claims. Rather it emerges from mostly implicit standards for aesthetic activities and experience. Examples of such standards would be the notions of articulation, depth, and intensity in Adorno's aesthetics (*AT* 187–92/279–87). Unlike Adorno, however, I do not restrict such standards to artistic matters, nor do I locate them in objects as such. They hold for aesthetic as well as artistic matters, and they apply to the intersubjective processes within which entities function as aesthetic signs. While the validity of such standards can always be discussed, questioned, contested, and ignored, the very struggle over them requires that people appeal to a principle of aesthetic validity.

I try to approximate this principle of aesthetic validity with the notion of imaginative cogency. Imaginative cogency is a historically emergent horizon of expectations concerning what makes aesthetic matters more or less worthwhile. It is "a horizon of aesthetic validity within which an intersubjective process unfolds, a horizon that encompasses the 'objects' of this process in their function as aesthetic signs." When people in modern Western societies evaluate the relative aesthetic merits of cultural products and events, such as jokes or metaphors or public celebrations, they "employ implicit standards of complexity, depth, and intensity, and the horizon of such standards is something like imaginative cogency."[27]

Hence, the autonomy of the aesthetic dimension, as it has taken shape in modern Western societies, lies both in the distinctiveness of imagination as a complex of intersubjective processes and in the inescapability of the normative horizon within which these processes occur. Such distinctiveness and inescapability do not isolate the aesthetic dimension. The aesthetic dimension is ubiquitous, showing up across the entire range of social institutions and cultural practices, and not simply in those which are artistic. Yet the aesthetic dimension is no more ubiquitous than, say, the economic dimension or the political dimension. For that reason, contra Menke, aesthetic ubiquity as such is not the source of a "crisis in reason." Rather, as

[27] Ibid., p. 64.

Adorno understood, the crisis arises from the failures of enlightenment rationality, *including aesthetic rationality*, to deliver what it promises, a topic to which I return in Chapters 2 and 4.

The relevance of aesthetic autonomy for understanding artistic autonomy becomes apparent when, following both Adorno and Habermas, one regards art as a historically institutionalized expert culture where aesthetic validity claims can be thematized. The development of Western societies has tended to isolate such expertise, to the impoverishment of both art and society. Adorno recognized this trend, but he turned it into a necessary condition for art's critical and utopian capacities. The tendency toward isolation has gone through three stages: first, a channeling of aesthetic efforts into the artworld, compensating for aesthetic impoverishment outside; second, a marginalizing of nonaesthetic concerns within the artworld, rendering the social significance of art opaque; third, an attack on aesthetic concerns within the artworld, such that "the last resistance to 'business as usual' threatens to disappear, albeit with a transgressive gesture." My attempt to link artistic autonomy with aesthetic autonomy arises from a refusal "to restrict the pursuit of aesthetic merit to the artworld, to make nonaesthetic concerns marginal to art, or to embrace the deaestheticization of either art or culture."[28]

Artistic Truth

Central to my own account of artistic autonomy, as it is to Adorno's, is a conception of artistic truth. My book *Artistic Truth* proposes a three-dimensional conception that builds on my account of the aesthetic dimension. Retaining but revising Adorno's idea of truth content (*Wahrheitsgehalt*), which Menke discards, I claim that artworks have import and that our interpretations of artistic import include an expectation concerning the artwork's integrity. We expect the artwork to be true with respect to its internal demands, including the requirement that it live up to more than its own internal demands. This expectation of integrity arises because artworks double the capacity of aesthetic signs to present nuances of meaning: artworks

[28] Ibid., p. 65.

"usually present something else by presenting themselves, and they present themselves in presenting something else."[29]

The expectation of integrity does not exhaust the truth of artworks, however, and it does not hold in the same way for artistic products and events that are not institutionally constituted as artworks. Here I have in mind the many varieties of folk art, popular art, and mass-mediated art that Adorno finds mostly incapable of being true. To expand the notion of artistic truth so that it encompasses such products and events, and also to demonstrate the wider reach of artistic truth for artworks proper, I introduce two other dimensions of artistic truth, namely, authenticity and significance. By "authenticity" I mean the expectation that art products and events "be true with respect to ... the experience or vision from which competent art making allows them to arise." By "significance" I mean the expectation that art products and events "be true with respect to a public's need for cultural presentations that are worthy of their engagement."[30] Hence, the truth of art products and events has as much to do with their contexts of production and use as it does with their internal configuration.

In all three dimensions – authenticity, integrity, and significance – artistic truth takes place as a process of imaginative disclosure. Through intersubjective processes of exploration, presentation, and creative interpretation that rely on suitable media and techniques, art discloses matters of vital importance that are hard to pin down. This disclosure occurs in the very contexts and configurations for which the expectations of artistic truth hold. Thus, when we expect an artwork to be authentic, integral, and significant, we expect it to be imaginatively disclosive of the experience from which it arises, of the artwork's own internal demands, and of a public's need for worthwhile cultural presentations.

This account of artistic truth implies, in turn, that autonomy need not isolate art from society. Just as the processes of imagination distinguish the aesthetic dimension from other dimensions of life and culture, so the expectation of imaginative disclosure distinguishes art – and not merely high art – from other fields of cultural endeavor. Just as the principle of aesthetic validity provides an

---

[29] Ibid., p. 129.   [30] Ibid., p. 128.

inescapable horizon for imagination, so artistic articulations of this principle (such as that art be original, challenging, and provocative) point toward an inescapable horizon for the imaginative disclosure that art provides. The latter horizon need not isolate art because the three dimensions of artistic truth together support occurrences of cultural orientation. By opening a window on the art maker's world, the achievement of authenticity also illuminates the interpreter's own personal world. Similarly, by challenging interpreters to recognize needs that go beyond personal preferences, the achievement of significance calls attention to a social world. In addition, by providing a configuration of import, artworks have "the capacity to wrench us free from both the personal and social worlds we already inhabit," to disclose what I call a "postsubjective world."[31] Accordingly, artworks elicit interpretations not only of themselves but also of the worlds to which they point, which exceed either the world of the artist or the world of the interpreter.

The artwork's capacity to disorient and reorient is central to Adorno's critically utopian aesthetics. It is why he insists on the autonomy of art. Menke's reconstruction turns this capacity into a transgressive aesthetic negativity that, in my view, holds little transformative potential. I have proposed instead that we retain a version of Adorno's "truth content" but recontextualize it within a more expansive notion of artistic truth, one that bursts Adorno's shrinking hothouse of authentic artworks. As a result, more of art can be considered autonomous than Adorno thought, and access to artistic truth need not be restricted to the professionals and cognoscenti of an expert culture. In principle, all who engage in the practices and institutions of imagination, whether these be artistic or not, have the potential to participate in artistic truth, provided they become conversant with the media of imagination at work in artistic products and events.

Multidimensional Autonomy

Having developed in tandem with the differentiation of an aesthetic dimension, the arts are societal sites for imaginative disclosure.

[31] Ibid., pp. 133–4.

They are autonomous in an explanatory sense: one cannot adequately explain their existence and function in society by reducing them to economic, political, psychological, or other factors. The arts are also autonomous in a justificatory sense: they make relatively unique contributions to human life and culture that are worthwhile and indispensable. What more would be required to make a viable case for the autonomy of the arts in the age of their global reproducibility?

Three things, at least: first, a sociological account of the ways in which art practices are constituted; second, a critical reconstruction of the economic underpinnings to the arts; and, third, a political theory that illuminates how arts organizations function in civil society and public spheres. In other words, a viable case requires attention to the multidimensional character of the arts and not simply to their aesthetic dimension. Let me indicate why the proposed case would constitute an account of autonomy *after* Adorno, and then suggest what it would include. Additional details will emerge in Chapter 5.

This chapter mentioned at the outset that Adorno's cultural theory and aesthetics employ three interconnected concepts of autonomy: the internal independence of authentic artworks, high culture's relative independence from the economic system, and the autonomy of political and moral agents. Accordingly, an account of autonomy after Adorno needs to go *beyond* aesthetic matters to take up topics in the sociology, economics, and politics of art. Moreover, contemporary forms of artistic practice such as "new genre public art" point in similar directions.[32] Very briefly, here is how the sociological, economic, and political elaboration of my account would go.

From a normative perspective, those contemporary art practices offer greatest promise which generate creative and critical dialogue via production of and participation in events, products, and experiences that are multifaceted, innovative, and attuned to current needs. Authentic works of art, in Adorno's sense, may or may

---

[32] See especially Suzanne Lacy, ed., *Mapping the Terrain: New Genre Public Art* (Seattle: Bay Press, 1995). I take up the topic of new genre public art in "Creative Border Crossing in New Public Culture," in *Literature and the Renewal of the Public Sphere*, ed. Susan VanZanten Gallagher and Mark D. Walhout (New York: St. Martin's, 2000), pp. 206–24.

not fit into such processes. In any case I shift the focus for justification from *works* as such to the quality of the *practices* in which artists and their collaborators and publics engage. This shift does not reduce artistic autonomy to the negativity of aesthetic experience, however.

Similarly, arts organizations are in a better position to pursue unique and worthwhile contributions if neither the corporate nor the government economy controls their economic viability and they sustain themselves within a not-for-profit countereconomy. The imperatives of money and administrative power are least conducive and most impervious to the sorts of dialogue and critical exploration that mark worthwhile art practices. Accordingly, arts organizations should do what they can to loosen the grip of corporations and governments on their own futures.

Still from a normative perspective, those artists and art publics are most likely to make unique and worthwhile contributions that build, maintain, and inhabit relatively independent public spheres. For, as Habermas has suggested in *Between Facts and Norms*, it is in autonomous public spheres that social solidarity and democratic processes can flourish. These, in turn, support intersubjective autonomy of the sort that autonomous art promotes and presupposes.

Now, admittedly, this abbreviated presentation raises the question whether my proposed account of art's autonomy retains negativity in art. The short answer is yes, but not in a manner that restricts it to aesthetic negativity. The negativity required has to do with developing alternative practices, economies, and public spaces that challenge dominant structures and systems. I am suggesting that what Danto calls "the end of art" does not simply free artists from art's "philosophical disenfranchisement" and from art's "need for constant self-revolutionization." It also frees artists to pursue "the basic needs to which art has always been responsive."[33] But I add an important caveat: this pursuit requires creative resistance to the ways in which basic needs are kept unfulfilled, especially among the poor, exploited, and marginalized in an increasingly globalized society.

[33] Arthur Danto, *The Philosophical Disenfranchisement of Art* (New York: Columbia University Press, 1986), p. xv.

With this caveat I point to subsequent chapters, which take up Adorno's emphasis on suffering and hope and ask about the possibilities for societal transformation. My dissatisfaction with Menke's reconstruction, and with much Habermasian criticism, stems from a desire to retain the utopian reach to Adorno's thought. Yet I accept elements of such reconstruction and criticism because I also find Adorno unable to articulate how societal transformation can actually occur. The aim of subsequent chapters is to understand this inability and to find ways to correct it, also with respect to how the arts can contribute to the transformation of society.

In fact, my own normative claims concerning the nonaesthetic dimensions to artistic autonomy would come to little if, descriptively, the recommended practices, organizations, and art publics either do not exist or are unlikely to exist. My research suggests that they do exist, and this receives confirmation from practical experience as the board president for a contemporary arts center. They are, however, beleaguered by commercial forces and government pressures, even as they try to break free from restrictive and outdated models of artistic autonomy. An aesthetically enriched and multidimensional account of autonomy can help remind such artists, arts organizations, and art publics that their contributions are uniquely worthwhile and cannot be replaced by the achievements of other professionals, businesses, or agencies. What they contribute cannot be limited, however, to either autonomous artworks or aesthetic negativity. Their autonomy should be dialogical rather than monadic; transformational rather than transgressive. It should be an autonomy *after* Adorno.

2

Metaphysics after Auschwitz

> It is astonishing how few traces of human suffering one notices in the history of philosophy.
>
> Adorno, *Negative Dialectics*[1]

Suffering and hope sustain Theodor W. Adorno's vision of philosophy. Not simply suffering, and not merely hope, but suffering and hope in their negative dialectical entwinement. And not simply Adorno's own philosophy, but any philosophy he would consider worth pursuing "after Auschwitz." His successors do not share his passions (*Leidenschaften*). In the polite language of Critical Theory after the communicative turn, they find Adorno's philosophy inappropriately "metaphysical" or "theological" or "utopian." Nor is this merely a generational difference of purely sociological interest. It goes to the heart of philosophy's tasks in contemporary society.

This chapter attempts to recover Adorno's passion without neglecting his dialectical precision. I first question the moves made

---

Parts of this chapter were presented in Halifax on May 31, 2003, at the symposium "Theodor W. Adorno (1903–1969)," cosponsored by the Canadian Philosophical Association and the Canadian Society for Aesthetics. I wish to thank Marie-Noëlle Ryan for her work in organizing the symposium, Albrecht Wellmer for his engaged interaction with the symposiasts, and all the participants for their instructive comments and questions. I also want to thank Deborah Cook, Ron Kuipers, Jon Short, and Bob Sweetman for illuminating comments on an earlier version of this chapter, and the students in my graduate seminar "Metaphysics after Auschwitz" for insightful conversations on Adorno's *Negative Dialectics*.

[1] Paraphrased from *ND* 153/156. Frequently I modify Ashton's translation.

in Albrecht Wellmer's critique of Adorno's "Meditations on Metaphysics."[2] Next, I explicate the Adornian themes of suffering and hope as ones that postmetaphysical philosophy mistakenly neglects. Then, through an exploration of two problems in Adorno's thematization of suffering and hope, I aim to indicate how these problems could be addressed without abandoning his social-philosophical project.

2.1 WELLMER'S POSTMETAPHYSICAL CRITIQUE

In an essay titled "Metaphysics at the Moment of Its Fall" (*EG* 183–201/204–23), Albrecht Wellmer interprets Adorno's project as attempting to develop "the notion of a way of thinking" that goes "beyond metaphysics" (*Begriff eines Denkens jenseits der Metaphysik*). This is an "aporetic" notion, Wellmer says: Adorno argues both that "the 'fall' of metaphysical ideas is irreversible" and that "the truth of metaphysics can only be grasped at the moment of its fall" (*EG* 183/204). Claiming that "a piece of unreconstructed ... metaphysics" illicitly circulates in Adorno's philosophy (*EG* 191/212), Wellmer proposes instead an approach to the concept of truth that frees it "from the confines of metaphysics" (*EG* 201/223).

Beyond Metaphysics?

Before we turn to details in Wellmer's provocative essay, I want first to question his general line of interpretation. On my reading, Adorno's meditations are not an effort to go "beyond metaphysics," as if they were taking a Kantian step beyond Kant. Rather they attempt to sublate (*aufheben*) metaphysics, to preserve and advance its "truth content," in the very process of criticizing both traditional metaphysics and Kantian and post-Kantian critiques of metaphysics. In other words, Adorno's is a Hegelian project, a negative *dialectic*, not a quasi-transcendental critique. Adorno's meditations mark the culmination to his materialist metacritique of German idealism.[3]

---

[2] These twelve meditations compose the third "model" in *Negative Dialectics* and the conclusion of the entire book.
[3] I derive this apt description from Simon Jarvis, *Adorno: A Critical Introduction* (New York: Routledge, 1998), pp. 148–74. See also my summary of Adorno's negative dialectic in the Appendix.

The way of thinking he seeks would not be "beyond metaphysics." It would be beyond the "identity compulsion" (*Identitätszwang* – ND 406/398) that disfigures not only German idealism but also modern capitalist societies. Yet even this movement "beyond" would occur by way of that which it negates. Perhaps it is Wellmer, not Adorno, who wishes to go "beyond metaphysics."

I also find it implausible to say that "the truth of metaphysics can only be grasped at the moment of its fall." Nor do I think that this is Adorno's position. To make this claim would be to assign to present conditions and to contemporary philosophy precisely that superior vantage point which Adorno rejects in Hegel's absolute idealism. Although Adorno also rejects Heidegger's regressively elevating the collapse of metaphysics into something metaphysical (*ND* 372/365), he would question the sanguine progressivism of philosophers who think they have a better grasp of things because modernization has moved them "beyond metaphysics."

To be sure, Adorno's project and Wellmer's general line of interpretation do seem to agree on the Weberian thesis that the collapse of metaphysical ideas is "irreversible." Like Habermas, whose critique of Adorno resembles Wellmer's, Adorno endorses "Kant's famous dictum that the critical path is the only one still open to us" (*CM* 7/461).[4] Yet even the secularization thesis, if I may call it that, needs refinement to be accurate to Adorno's project. For the point of his meditations is not simply to continue the trajectory of modernization but to prepare the way, or at least to hold open the door, for a fundamental transformation (*Umschlag*) in the society and philosophy that have taken this trajectory. Adorno's "Self-Reflection of the Dialectic" (*ND* 405–8/397–400), the last in his "Meditations on Metaphysics," points in this direction. Just prior to that section, he says "the world's course is not absolutely conclusive [*geschlossen*]." That is why, although metaphysics cannot be resurrected, perhaps it would arise "only with the actualization of what has been thought in its sign" (*ND* 404/396). This suggests that the collapse of metaphysics is not permanent – indeed, that the truth of metaphysics could only be "grasped" were both society and philosophy to undergo a fundamental transformation. Precisely such futurity, tentative though it

---

[4] Cf. Habermas, *PP* 14/29–30.

be, gives Wellmer pause. But to label this element a "piece of unreconstructed metaphysics" whereby Adorno "precritically undercuts Kant" (*EG* 190–1/211–12) is to miss the truth to Adorno's metacritique of metaphysics.

Materialist Metacritique

The truth to Adorno's metacritique lies in the insight that metaphysical questions about life, death, and immortality are not primarily epistemological in either intent or effect. Rather they are social-philosophical questions that encompass politics, ethics, and religion. Moreover, they are indexed to the historical conditions in which they arise and to the larger historical process to which they contribute. As a critical follower of Marx, Adorno insists that this process has a material dynamic from which not even philosophy can be removed. All of this is implied by Adorno's transition into the "Meditations on Metaphysics." The preceding model on "World Spirit and Natural History" ends with "the transmutation of metaphysics into history" (*ND* 360/353), and with the contemporary impossibility of recollecting transcendence other than by way of the transient (*Vergängnis, das Vergängliche*).

That sets the stage for "After Auschwitz," the opening section in "Meditations on Metaphysics." Unlike traditional metaphysics, contemporary philosophy can no longer tie truth to immutability, Adorno says, nor regard what changes as semblance (*Schein*), nor maintain a separation between eternal ideas and temporal phenomena. Instead, dialectical philosophy, secularizing a mystical impulse, considers the contemporary historical world to be relevant for "transcendence," or at least for the position consciousness takes toward traditionally metaphysical questions (*ND* 361/354).

Adorno's emphasis on transience calls Hegelian attention to the temporal and historical character of metaphysical ideas. But it also gives a Marxian nod to their proleptic character: they point toward change or, more strongly, toward transformation that has not yet occurred and whose occurrence is not yet impossible.[5] They cannot

---

[5] On the eschatological character of Adorno's philosophy of history (*Geschichtsphilosophie*), see Michael Theunissen, "Negativität bei Adorno," in *Adorno-Konferenz 1983*, ed. Ludwig von Friedeburg and Jürgen Habermas (Frankfurt am Main:

be fully proleptic, however. The idea of truth, for example, cannot simply represent or assume utopia as if it presently exists or as if it were already accomplished. Any tendency toward proleptic assurance would make metaphysical ideas unbearable "after Auschwitz," when every attempt to derive future-oriented meaning from current existence seems "sanctimonious" and disrespectful to the victims (*ND* 361/355). Auschwitz has shattered the basis for unifying "speculative metaphysical thought" and "experience" (*ND* 362/354). Whereas Kant asked epistemologically whether and how metaphysics is still possible as a science, Adorno will ask historico-philosophically whether and how "metaphysical experience" is still possible (*ND* 372/364–5), not as a science, but as a basis for philosophy that has a society-transforming intent.

It is so, as Wellmer claims, that Adorno considers "truth" to be "supreme [*die oberste*] among metaphysical ideas" (*ND* 401/394). It is also so that defending the idea of truth is "what really concerns him" (*EG* 187/208) in his "Meditations on Metaphysics." Contrary to Wellmer's interpretation, however, Adorno's idea of truth does not have primarily epistemological intent and effect. Nor does Adorno's defense of the idea aim simply to shore up philosophy itself. If we take seriously Adorno's insistence on the sociohistorical character of metaphysical questions and on the material dynamic of history itself, then we should regard his idea of truth as having primarily a social-philosophical character.

In this sense, Adorno pushes Kant's "postulates of pure practical reason" beyond their inherently dualistic framework into a transformational social philosophy.[6] Kant asked which ideas about the soul, the world, and God are needed in order for human beings to continue striving for the highest good, as we are morally obligated to do, despite the hindrances posed by our finitude and

---

Suhrkamp, 1983), pp. 41–65, especially pp. 53–7. Theunissen suggests that in Adorno's eschatology apocalypse is even more fundamental than prolepsis. Like Wellmer, although with less animus against traditional metaphysics, Theunissen argues that the aporias in Adorno's negative dialectic drive him into the arms of theology. For a summary and response to Theunissen's essay, see Jarvis, *Adorno*, pp. 211–16.

[6] Strictly speaking, the Kantian framework is not dualistic but a type of monism in which the mental, especially the rational, has priority. The Dutch philosopher Dirk Vollenhoven would call it a type of "ennoetism."

corporeality.⁷ Adorno, by contrast, asks what transformations in society and philosophy would be both possible and required in order for needless suffering to end. The truth content of metaphysics, including the "metaphysical" idea of truth itself, lies in keeping this question open.

Wellmer touches on this when he says Adorno replaces Kant's "We cannot know [the absolute]" with "we do not know it *yet*" (*EG* 190/ 211). But Wellmer pursues this in a primarily epistemological direction. He reads Adorno as precritically concluding "from the historical character of our forms of thought and intuition" that "the [a]bsolute as reconciliation... could become a historical reality." This conclusion is philosophically naive, Wellmer suggests, because "we can already know *now* that we cannot anticipate as real that which we cannot even consistently *conceive* [*denken*] as real" (*EG* 190–1/212). The only potential resolution Wellmer credits for "the aporetic relationship between the necessity and impossibility of metaphysics" is epistemological, not societal: philosophy could make some conceptual advances or "reformulate the questions that apparently permitted only aporetic answers" (*EG* 190/211).

From an Adornian perspective, Wellmer's objections are problematic in several respects.⁸ First, Wellmer's appeal to what "we can already know *now*" assigns a constant and unchanging validity to "our" forms of consciousness. That is precisely what Adorno rejects in Kant, on critical and not precritical grounds. Unless one rejects outright the Hegelian critique of Kant that Adorno reworks in his own materialist way – and Wellmer does not appear to reject it – one cannot simply appeal to what we can know now without begging the question of how present knowledge is historically mediated. Second,

---

⁷ See especially part 1, book II in Kant's *Critique of Practical Reason*, in Immanuel Kant, *Practical Philosophy*, trans. and ed. Mary J. Gregor, general introduction by Allen Wood (Cambridge: Cambridge University Press, 1996), pp. 226–58; *AK* 5:107–48.

⁸ I agree with Simon Jarvis's general point that, because Adorno does not think metaphysical experience can be eliminated from thought, Adorno does not straightforwardly endorse "the very distinction between pre-critical and critical thinking which Wellmer, like almost all second-generation critical theorists, takes as a benchmark." But Jarvis concedes too easily Wellmer's claim that we cannot anticipate as real what we cannot consistently think as real. Unlike Jarvis, I do not think this claim "would constitute a decisive objection to Adorno's account of his relation to metaphysics" (Jarvis, *Adorno*, p. 209), for reasons I explain in the text.

Wellmer's emphasis on what "we" cannot "consistently *conceive* as real" puts a premium on logical consistency without acknowledging that such privileging of "identity thinking" is exactly what Adorno himself links with the underlying societal principle – the exchange principle – that would have to change. Adorno does not reject logical consistency, of course, but he does suggest its insufficiency. Third, Wellmer seems to think that philosophical advances in matters metaphysical can occur independently from significant sociohistorical transformation. But this is to assume that the issues in dispute are primarily conceptual and logical problems rather than sociohistorical dilemmas. Although many contemporary philosophers might share this assumption, it is not Adorno's assumption, nor, I dare say, was it Marx's, Hegel's, or even Kant's. To reduce metaphysical issues to conceptual problems would be to misread the critical tradition that Adorno's alleged "piece of unreconstructed metaphysics" purportedly falls behind.

Priority of the Object

Even more problematic, however, is Wellmer's construal of the relation between historical mediation and "the absolute." His construal omits Adorno's well-known emphasis on the "priority of the object" (*Vorrang des Objekts*). Admittedly, this is a protean and diffuse emphasis. But its importance can hardly be overlooked. Adorno himself says *Negative Dialectics* aims to complete his lifelong task as a philosopher: "to use the strength of the [epistemic] subject to break through the deception [*Trug*] of constitutive subjectivity" (*ND* xx/10).

Let me mention two of the many ramifications to the priority of the object. The first is that Adorno's stress on the societal preformation of consciousness replaces idealist notions of constitutive subjectivity. Such societal preformation holds for both individual consciousness and intersubjective communicative action. Accordingly, the idea of truth cannot but be societal in its content and implications. That is one sense in which the object must have priority: the societal content and implications of ideas and practices must always come to the fore.

This helps explain why Adorno uses the subjunctive ("*das Absolute wäre*") when, in opposition to Hegel, but also to Kant, the final meditation "identifies" the absolute. Here Adorno asks whether metaphysics, as knowledge of the absolute, is possible without

Hegel's presumption of absolute knowledge. The dilemma is this. If the absolute is conceived dialectically, then dialectical thought poses as a form of absolute knowledge, contrary to the idea of negative dialectics. But if the absolute is conceived as being completely incommensurable with dialectics, then one resorts to a double truth, contrary to the idea of truth. Metaphysics depends upon whether thought can get out of this aporia "without subterfuge" (*ND* 406/397). Adorno's response is to have dialectics turn against itself "in a final movement." Dialectics must turn against its own unavoidable tendency to absolutize the concept and conceptual identity. In this turn it absorbs the strength of the identity-governed "immanence context" in order to break out of that context from within. Using the means of logic, "dialectics grasps the coercive character [*Zwangscharakter*] of logic, hoping that this may yield. For that coerciveness is itself the mythical semblance [*Schein*], the forced identity. But the absolute, as metaphysics envisions it, would be the nonidentical that would not emerge until after the compulsion of identity [*Identitätszwang*] would dissolve" (*ND* 406/398). The rest of *Negative Dialectics* makes clear that this compulsion would not dissolve unless society itself underwent fundamental transformation.

The second ramification to the priority of the object concerns the materiality of culture and consciousness – of *Geist*, in Adorno's Hegelian language. In the passage from which the title of Wellmer's essay derives – the last lines of *Negative Dialectics* – Adorno insists that his sublation of metaphysics into "micrology" requires thought to proceed from the material need (*das Bedürfnis – zunächst die Lebensnot*) contained in thought as a mode of conduct (*ein Verhalten*) or action (*als Tun*). Thought must proceed from such need and must sublate it. The need within thought, which amounts to the nonidentical there, attunes thought to that which thought cannot know so long as it follows the compulsion to identify (*Identitätszwang*). It is because thought cannot rid itself of all materiality that Adorno retains hope for a society in which material needs would be met and suffering would diminish.[9]

---

[9] For a lengthier discussion of Adorno's emphasis on corporeality, from a feminist perspective, see Lisa Yun Lee, *Dialectics of the Body: Corporeality in the Philosophy of T. W. Adorno* (New York: Routledge, 2005).

Permit me to quote Adorno at some length, in my own translation:

> The need in thinking [*Denken*] desires that thought occur. It [i.e., the need in thinking] demands its own negation by thinking [*Denken*], must disappear in thinking to be really satisfied; and it survives in this negation, represents [*vertritt*] in the innermost cell of thought [*des Gedankens*] what is not like thought [*nicht seinesgleichen*]. The smallest inner-worldly stirrings [*Züge*] would have relevance for the absolute; for the micrological view cracks the shells of what, measured by the subsuming cover concept, is helplessly isolated, and explodes its identity, the delusion that it is a mere specimen. Such thought has solidarity with metaphysics in the moment of its collapse. (*ND* 408/399–400)

Although one could question the moves Adorno makes here, it is simply not so that he straightforwardly and problematically concludes from the historical mediation of thought to the possibility of the absolute becoming actual. Nor does he resort to an "aesthetic" escape hatch, as many Habermasians claim. Wellmer's criticisms retain their plausibility only so long as one brackets the two ramifications I have mentioned. Conversely, the plausibility of what Wellmer labels a "materialistic appropriation of theology" (*EG* 191/212) depends upon Adorno's twofold stance that society preforms consciousness and that material need propels thought.

One other ramification to the priority of the object deserves mention. This one would remove what Wellmer sees as "an insoluble conflict between materialist and metaphysical (i.e., theological) motifs" in Adorno's approach. Wellmer separates "messianic hope" or "the hope of salvation" from the transformation of society or "the transfiguration of historical reality." He poses this as a dilemma that Adorno allegedly cannot resolve: "If the hope of salvation were to be fulfilled in history, it would not be the hope of salvation that was fulfilled (but rather that of a fulfilled life). On the other hand, if what was fulfilled were really the hope of salvation, this would still not signify a new condition of *history*" (*EG* 191/212). This posing of the dilemma assumes the traditional otherworldly eschatology already questioned by the Jewish mystics and Christian socialists with whom Adorno was conversant.[10] The dilemma does not accord with the

---

[10] Here I have in mind especially the scholarship of Gershom Scholem, with whom Adorno edited a two-volume collection of Walter Benjamin's letters, and Paul

most productive eschatological theologies of the past three decades, such as the "liberation theology" of Gustavo Gutiérrez and Jürgen Moltmann's "theology of hope." Nor is the separation of salvation from a historical transformation of society demanded by the sacred writings of Judaism and Christianity, where imagery of earthly "shalom" and societal reconciliation abounds. Rather than posing a dilemma, Adorno's interpretation of the messianic condition as societal, historical, and material complements theologies that do not equate salvation with the release of individual souls into a state of disembodied immortality.[11]

Adorno shows his awareness of such "materialist" potential within religious traditions in a passage on resurrection that Wellmer quotes, but without exploring its social-philosophical implications. Compared with speculative metaphysics, Adorno writes, Christian dogmatics was "metaphysically more consistent – more enlightened, if you will" when it connected the awakening of souls with "the

---

Tillich, who supervised Adorno's *Habilitationsschrift* on Kierkegaard and remained a lifelong friend. Habermas's "Gershom Scholem: The Torah in Disguise" (*PP* 201–13) gives an indirect indication of how Jewish mysticism might resonate with Adorno's negative dialectic. See also Habermas's "The German Idealism of the Jewish Philosophers" in the same volume (*PP* 21–43/37–66). Mauro Bozzetti touches on Adorno's "close kinship to the speculations of modern Hebrew philosophy" in "Hegel on Trial," in *Adorno: A Critical Reader*, ed. Nigel Gibson and Andrew Rubin (Oxford: Blackwell, 2002), pp. 292–311, especially pp. 300–4. For Adorno's personal reminiscences about Paul Tillich, see pp. 24–38 within "Erinnerungen an Paul Tillich," in *Werk und Wirken Paul Tillichs: Ein Gedenkbuch* (Stuttgart: Evangelisches Verlagswerk, 1967). This is the transcript of a radio program broadcast on August 21, 1966. It is interesting to note, as Rolf Tiedemann points out in his "Editor's Afterword" to Adorno's posthumous lectures on metaphysics, that, when writing his "Meditations on Metaphysics," Adorno asked to borrow the third volume to Tillich's *Systematische Theologie*. See Theodor W. Adorno, *Metaphysics: Concept and Problems* (1965), ed. Rolf Tiedemann, trans. Edmund Jephcott (Stanford: Stanford University Press, 2000), p. 194. I am indebted to Matt Klaassen for this reference.

[11] I do not mean to suggest, however, that Adorno shares the more affirmative vision of these theologies. For a nuanced reading of Adorno's relevance for "academic theology" that regards his emphasis on negativity as a corrective to the "meaning-optimism" of postmetaphysical theories, see Mattias Martinson, *Perseverance without Doctrine: Adorno, Self-Critique, and the Ends of Academic Theology* (Frankfurt am Main: Peter Lang, 2000). See also the attempts to retrieve the theme of hope from both theological misconstrual and post-theological neglect in Miroslav Volf and William Katerberg, eds., *The Future of Hope: Christian Tradition amid Modernity and Postmodernity* (Grand Rapids, Mich.: Eerdmans, 2004).

resurrection of the flesh." So, too, "hope means corporeal resurrection" and loses its best element when it is spiritualized (*ND* 401/393).[12] If, with Adorno, and in line with productive theologies of recent decades, one interprets "the hope of salvation" as a hope for fundamental transformation in society, then the apparent dilemma of either salvation or historical transfiguration disappears. And if, like Adorno, one accepts "the transmutation of metaphysics into history" and does not hold them in classical opposition, then one can consider Adorno's approach "metaphysically more consistent – more enlightened, if you will" than a postmetaphysical critique that relegates issues of suffering and hope to an outmoded "religion" and detaches questions of truth from "the resurrection of the flesh."

2.2 SUFFERING, HOPE, AND SOCIETAL EVIL

Wellmer's objections signal a gap between Adorno's vision of philosophy and the reigning paradigm today. Contemporary strands of Western philosophy converge in the claim that philosophy must be "postmetaphysical." This requirement is considered a matter of both historical necessity and disciplinary integrity. Surely, contemporary Western philosophers seem to say, no up-to-date and self-respecting philosophy can be metaphysical. This postmetaphysical presumption affects philosophy as a whole: its tasks, its relationships to other forms of inquiry and practice, and its political relevance in contemporary society. Adorno's vision challenges contemporary philosophy's self-understanding. According to his "Meditations on Metaphysics," philosophy must *incorporate* "metaphysical experience" rather than "go beyond" metaphysics. Otherwise, philosophy cannot remain unswervingly self-critical, engage in thorough social critique, and hold open historical alternatives to contemporary society.[13]

---

[12] Wellmer quotes this passage in *EG* 186/206–7.
[13] In conversation Albrecht Wellmer has raised the question whether abandoning the "metaphysical" side of Adorno's project actually *necessitates* giving up a critique of society as a whole. I am not ready to argue for a necessary relation here, but I do not regard as mere coincidence the tendency for both developments to occur together in much of Habermasian Critical Theory. Traditionally, some vision of "the good," however partial and implicit, was presupposed when theorists tried to identify a principle or dynamic that unifies societal ills into societal evil. Against this

If Adorno is right, then his own metacritique of metaphysics merits a critical retrieval. The themes of suffering and hope are central to such a retrieval. This means, to begin with, that one cannot ignore how Adorno's metacritique positions itself "after Auschwitz." "To write poetry after Auschwitz is barbaric," Adorno once said. Revising Adorno's famous claim, which he himself revisited, one could argue from his "Meditations on Metaphysics" that to write suffering and hope out of philosophy is barbaric.

Philosophical Avoidance

What does this positioning mean for Adorno's project? I do not think his metacritique of metaphysics is simply an attempt to see Nazi genocide as "fulfilling the logic of disenchantment"[14] or as "the moment of accomplishment and self-destruction of [the European] Enlightenment" (Wellmer, *EG* 183/204). As Espen Hammer suggests, it is also Adorno's attempt to work out the implications of "extreme evil" for a society and a philosophical trajectory that resist or reject anything outside their confines. This is why, in Hammer's words, "Adorno's challenge is that post-metaphysical philosophy is not post-metaphysical enough. By discarding immutability and transcendence but without questioning the claim for radical immanence itself, a claim Auschwitz irretrievably has shown to have failed, it is not sufficiently alive to ways of thinking transcendence that would escape the charge of being affirmative."[15]

Adorno suggests many ways in which contemporary philosophy avoids this challenge. One is the reluctance of philosophy to be radically self-critical. To be true today, he says, thought must also "think against itself." Otherwise it will be mere background music that covers up "perennating suffering" (*das perennierende Leiden*), which has as much right to be expressed as the tortured victim has to

---

backdrop, it is difficult to be "beyond good" without being "beyond evil" as well, as Nietzsche suggested.

[14] J. M. Bernstein, *Adorno: Disenchantment and Ethics* (Cambridge: Cambridge University Press, 2001), p. 383.

[15] Espen Hammer, "Adorno and Extreme Evil," *Philosophy and Social Criticism* 26, no. 4 (2000): 75–93; quotation from p. 79. I use the term "societal evil" as a near equivalent for what Hammer labels "extreme or radical moral evil."

scream (*ND* 362/355, 365/358). A complementary tendency is the unacknowledged continuation of traditional metaphysical separations between body and soul. This ratifies a societal division between physical and mental labor and supports inattention to "questions of material existence" (*ND* 366/358). A third indication of avoidance is philosophy's inability to come to grips with the experience of death, despite an ideological "death metaphysics" stemming from Heidegger's *Being and Time* (*ND* 368–73/361–6). Yet another sign lies in a reluctance to think through the tension between a promise of fulfillment (*Glück*) and the experience of waiting in vain (*vergebliches Warten*) for fundamental transformation to occur (*ND* 373–5/366–8). Add to these tendencies various falsely posed questions about "the meaning of life" (*ND* 376–81/369–74), and one has ample evidence that, on the road to becoming postmetaphysical, contemporary philosophy has largely surrendered the task of explicating "metaphysical experience" in the face of societal evil. It is against the backdrop of these tendencies that Adorno undertakes his materialist metacritique of Kant and Hegel.[16]

In Adorno's terms, to surrender the explication of metaphysical experience would also mean that philosophy has largely abandoned the project of a comprehensive critique of society. This would imply in turn that, despite increasingly sophisticated discussions of politics, ethics, and religion, contemporary philosophy has become unethical: it has failed to take seriously enough what Adorno proposes as "a new categorical imperative," namely, that human beings [including philosophers] so "arrange their thought and action that Auschwitz would not repeat itself, [that] nothing similar would happen." Anticipating the objections of philosophers for whom argumentation trumps experience even in the face of unspeakable suffering, Adorno adds: "This imperative is as resistant to justification [*Begründung*] as the givenness of the Kantian [categorical imperative] once was" (*ND* 365/358).[17] A discursive treatment of the new imperative would be an

---

[16] See especially meditations 6–12 (*ND* 381–408/374–400).
[17] Adorno's essay "Education after Auschwitz" opens in a similar way: "The premier demand upon all education is that Auschwitz not happen again. Its priority before any other requirement is such that I believe I need not and should not justify it.... To justify it would be monstrous in the face of the monstrosity that took place." *CM* 191/674.

"outrage" (*Frevel*), he says, violating not a human principle or divine law but the moment of ethical excess (*das Moment des Hinzutretenden am Sittlichen*) that the imperative lets one feel corporeally. The corporeal feeling is an abhorrence of physical pain, an abhorrence-become-practical toward the unbearable physical pain to which individuals are exposed. Morality survives, he says, in this "materialistic motive" (*ND* 365/358). The same goes for metaphysics, once corporeal suffering in Nazi concentration camps burned away any comfort intellectual culture could provide.

To Let Suffering Speak

For Adorno, the key to avoiding philosophy's avoidance of societal evil is also a key to philosophy's pursuit of truth: "The need to let suffering speak [*beredt werden zu lassen*] is a condition of all truth. For suffering is objectivity that weighs upon the subject..." (*ND* 17–18/29). The need to express suffering is a primary motivation for Adorno's critique of identitarian thought, his insistence on nonidentity, his emphasis on conceptualizing the nonconceptual, and the stress his philosophy places on linguistic presentation and conceptual constellations. Although his articulation of these themes has considerable relevance for epistemology, philosophy of language, and philosophy of science, his motivation for discussing them lies beyond the boundaries of such philosophical subdisciplines. It lies in a "philosophical experience" where suffering and the need to express it are as unavoidable as they are compelling. For suffering defies discursive treatment, yet it calls for conceptual comprehension if philosophy is to resist both forgetting and perpetuating suffering. Although such comprehension will not render suffering conceptual, it will seek to understand its societal causes and social significance.

Traditional metaphysics informed by Hebraic wisdom literature asked why good people suffer. Adorno asks why, in a society that has the means to eliminate poverty, hunger, and economic exploitation, suffering continues unabated and even takes the forms of genocide and mass destruction. If this is a "metaphysical" question, then it is also a central question of social critique. To avoid it would be to give up philosophy's pursuit of truth and to seal its political irrelevance. So the issue Adorno poses for contemporary philosophy is twofold: whether societal evil is inevitable, and whether a good society is historically possible. He wants to reject such inevitability, while looking societal evil squarely in the face. And he wants to affirm the historical possibility of a good society, while demolishing premature affirmations of the goodness of contemporary society. In each case the "goodness" or "truth" of society is indexed to both the remembrance and the elimination of suffering.

This double gesture provides impetus for Adorno's critical retrieval of Kant's postulates of pure practical reason. Just as Kant's critical resolution (*Aufhebung*) of the antinomy of practical reason hinges on his discovering a type of happiness "caused" by virtue – namely, the "intellectual contentment" one feels as a result of virtuously obeying the moral law[18] – so Adorno's critical retrieval of Kantian ethics hinges on his recognizing a type of resistance to societal evil occasioned by a corporeal feeling of abhorrence toward suffering. Just as Kant finds it necessary for human beings to postulate the immortality of the soul in order to strive for a state in which their happiness and virtue would coincide,[19] so Adorno finds it necessary to maintain the historical possibility of a good society in order for human beings to strive for a world where material needs are satisfied and needless suffering ends. In shifting the emphasis from pursuing moral goodness to resisting societal evil, Adorno rejects the separation between body and soul that sustains Kant's conception of the highest good. Adorno also refuses to isolate personal moral goodness from the society in which individuals are constituted as individuals.

These differences help explain why Adorno says the secret to Kant's philosophy is "the incomprehensibility [*Unausdenkbarkeit*] of despair" (*ND* 385/378). By advocating the practical necessity of postulating the immortality of the soul, for example, Kant recognizes in his own ideologically distorted way that there would be no genuine prospect of a good society if death had the final word. This is why, in an earlier passage, Adorno says the thought that death is final and absolute is "impossible to consider" (*unausdenkbar*), just as impossible to consider as the idea of immortality is. Desire (*die Lust*) resists transience (*Vergängnis*). So does thought itself. If death were absolute,

[18] Kant, *Critique of Practical Reason*, pp. 234–6; *AK* 5:117–19.
[19] Ibid., pp. 238–9; *AK* 5:122–4.

then everything would be absolutely nothing (*überhaupt nichts*), every thought would be empty (*ins Leere gedacht*), and no thought could be thought in a true fashion (*keiner lässt mit Wahrheit irgend sich denken*). "For it is a moment of truth that, along with its temporal core, truth should last; without any duration there would be no truth, [for] absolute death would swallow up its final trace" (*ND* 371/364).

But the incomprehensibility of death and despair has relevance for philosophy only if philosophy does not refuse the attempt to comprehend them. The insistence on this "speculative moment" (cf. *ND* 15–18/27–9) runs directly contrary to a society where, according to Adorno, individual and collective self-preservation have become a structural obsession. To resist such a society, philosophy must maintain a moment of independence from self-preservative business as usual. It must insist on the priority of the object (*ND* 388–90/381–2). Accordingly, Kant's concept of the intelligible world, as something necessarily postulated but not necessarily existent, serves to point beyond the immanence of self-preservation. "The gesture of hope is to let go what the subject wants to cling to, what [the subject] expects will endure. The intelligible... could only be thought negatively" (*ND* 392/384). And that, paradoxically, would make the intelligible an "appearance" (*Erscheinung*) – what the nonidentical discloses to the finite spirit (*was das dem endlichen Geist Verborgene diesem zukehrt*),[20] what the finite spirit is compelled to think and deforms. "The concept of the intelligible is the self-negation of finite spirit" (*ND* 392/384). The concept of the intelligible registers how what merely exists becomes aware of its insufficiency – becomes aware of this in spirit. Taking leave of such self-enclosed (and insufficient) existence gives rise to that speculative moment in which spirit separates from its own principle of self-preservation. Spirit transcends itself in self-negation. To be spirit, spirit must know that it is not exhausted in finite existence. That is why spirit thinks what lies beyond it in the concept of an intelligible world.

On Adorno's interpretation, then, the metaphysical experience inspiring Kant's philosophy is the negation of the finite that finitude

---

[20] This is one of the passages where Adorno's mostly unacknowledged proximity to Heidegger is readily apparent. Even the terminology – *das Verborgene*! – is Heideggerian, and it is not used ironically or caustically.

requires. The "intelligible" points to spirit's moment of independence from what exists, the moment spirit attains when it insists on the nonidentical as distinct from spirit. Metaphysics has an inconspicuous possibility in spirit's "moment of transcendent objectivity." "The concept of the intelligible realm would be the concept of something that does not exist and yet is not simply nonexistent [*etwas, was nicht ist und doch nicht nur nicht ist*]" (*ND* 393/385). Yet we must not conclude from this concept that the intelligible already actually exists (as happens in ontological proofs for God's existence, which Kant destroyed). Stringent critique of the insufficiency of what exists does not remove this insufficiency.[21]

"Weh Spricht: Vergeh"

At this point the Adornian emphasis on suffering turns into an emphasis on hope. Or, rather, the philosophical expression of suffering receives articulation as an expectation of its removal. For Adorno, the material need within thought propels thought toward the idea of a fundamentally transformed world within which thought itself would be thoroughly transformed. Although the need is itself societally mediated, it is neither absorbed nor satisfied by its societal mediation. Suffering and hope are complementary manifestations of this unmet need. Both are ineliminable from thought: "*Weh spricht: vergeh*" (*ND* 203/203).

[21] A remark by Albrecht Wellmer at the Adorno symposium in Halifax leads me to think that the question of human finitude separates his reading of Adorno from my own. Both Wellmer and I want philosophy to take human finitude seriously, and we find this impulse in Adorno too. Whereas Wellmer's reading emphasizes the *inescapability* of finitude, my reading stresses the *insufficiency* of finitude. I think both themes are prominent in Adorno's "Meditations on Metaphysics." Their interlacing makes Adorno's metacritique so provocative. If *accepting* the inescapability of human finitude were a hallmark of postmetaphysical philosophy, however, then I would argue that neither Adorno's project nor my own critical retrieval is postmetaphysical. In opposition to Habermas's vision of postmetaphysical philosophy, this would mean refusing either to accept that "the nonobjective whole of a concrete lifeworld... evades the grasp of theoretical objectification" or to let all "explosive experiences of the extraordinary" safely migrate out of philosophy into autonomous art and into a subrational religion that provides "normalizing intercourse with the extraordinary." See Jürgen Habermas, *Postmetaphysical Thinking: Philosophical Essays*, trans. William Mark Hohengarten (Cambridge, Mass.: MIT Press, 1992), pp. 50–1.

Adorno's sources of hope are mixed and scattered. They include the import of humanist culture, especially Kant and Beethoven (*ND* 397/389–90); the unredeemed promises of religious traditions (e.g., *ND* 401/393); the transience of everything cultural and societal; and the advanced state of productive forces in contemporary society. His object of hope is a future society "without unfulfilled needs" (*ohne Lebensnot* – *ND* 398/390). This is the import, I take it, of Adorno's saying "hope means corporeal resurrection" (*ND* 401/393).

Both the sources and the object of hope come together in Adorno's appeal to an experience that attends speculative thought. He appeals to the experience that thought which "does not decapitate itself" flows into the idea of a world where "not only extant suffering would be abolished but also suffering that is irrevocably past would be revoked." It is the experience of having "all thoughts converge in the concept of something that would be different" from today's unspeakable world (*ND* 403/395). Contemporary society is both "worse than hell and better," Adorno says. Worse, because there is no escaping it. Better, because the world's "disturbed and damaged" course cannot be construed as purely meaningless and blind. The world's course resists the desperate attempt "to posit despair as an absolute." No matter how weak the historical traces of the other, no matter how disfigured all happiness due to its revocability, the promises of the other, though broken ever again, still pervade what exists, in the breaks that resist identity. "Every happiness is a fragment of the entire happiness that is denied to human beings and that they deny themselves" (*ND* 404/396).

Like the experience of suffering, then, the experience of hope has an ineliminable materiality. Just as suffering is how objectivity weighs upon the subject, so hope arises because "something in actuality [*in der Sache*]" presses toward "the humanly promised other of history" (*ND* 404/396). Both suffering and hope are implied, it seems to me, when Adorno describes his own micrological "metaphysics" as "a legible constellation of what exists [*von Seiendem*]," receiving its material (*Stoff*) from what exists, but configuring the elements to form a script (*ND* 407/399). Both suffering and hope are implied when Adorno says such micrological thought must proceed from the material need sublated within itself. Such thought, materially motivated and materially attentive, would have solidarity with metaphysics, he says, "in the moment of its collapse" (*ND* 408/400).

2.3 DISPLACED OBJECT

Earlier I raised several objections to Albrecht Wellmer's critique of Adorno. Wellmer claims we cannot anticipate as real what we cannot consistently conceive as real. Adorno, by contrast, claims that what we inescapably experience as real compels philosophy to criticize its own restrictions on what can be conceived, so as not to foreclose upon the historical possibility of a fundamental transformation in society. Both suffering and hope, materially rooted and societally mediated, let us experience real societal evil that calls for resistance. Just as to ignore suffering would intellectualize the experience of societal evil, so abandoning hope would remove the horizon within which societal matters can be recognized as evil to be resisted. Social philosophy would then be left with "pathologies," "crisis tendencies," "anomalies," and "problems" whose actual removal, though dramatic, would not require a radical transformation of society as a whole. Having said this in Adorno's defense, I also want to state two reservations, with the aim of critically retrieving his social philosophy. My reservations have to do, first, with his privileging of "philosophical experience" and, second, with his objectification of hope. Let me discuss each in turn.

Privilege of Experience

By singling out the need to express suffering as a condition of all truth, Adorno puts his own philosophy in a precarious position. On the one hand, he seems to cut off debate about the nature of truth, contrary to his own emphasis on dialectic and the concept. On the other hand, he appears to beg numerous questions about whose suffering and what manner of suffering need to be expressed, contrary to his emphasis on micrology and the nonidentical. Adorno himself acknowledges this precariousness when the introduction to *Negative Dialectics* discusses a "privilege of experience" (*ND* 40–2/50–3). There he mentions two objections to his notion of philosophical experience: (1) that such experience cannot be intersubjectively tested, and (2) that making philosophical experience a condition of knowledge is elitist and undemocratic. His reply touches on the first objection and addresses the second. He (1) *suggests* that intersubjective testing might not be so decisive in any case, because the tests take place under distorting societal conditions; and he (2) *claims* that the administered world does not give everyone an equal capacity to engage in critique. Societal conditions have made many people incapable of experience in an emphatic sense. Under such conditions, he says, to construe truth as the will of the majority would invoke democracy to deceive everyone about what they need. Those who (like Adorno) have had the undeserved luck to resist prevailing norms have a moral obligation to speak on behalf of others and to express what most of them either cannot see or realistically refuse to see.

Perhaps we can call this a charismatic or prophetic vision of the philosopher's task in society. Adorno connects this prophetic vision with what Wellmer and Habermas would call an esoteric view of truth. What makes something true, says Adorno, is not its being immediately communicable to everyone. We must neither confuse communication with what is known nor rank it higher. "Today every step toward communication sells out the truth and makes it false" (*ND* 41/51–2). Although truth requires subjective mediation, it is its own index. Truth loses its supposedly privileged character by not making special pleas for the experiences to which truth is indebted and by entering contexts of justification (*Begründungszusammenhänge*) that bear it out or establish its inadequacies.

I do not want to deny that contemporary societal conditions render many people incapable of experience in an emphatic sense. Neither do I wish to make communicability a decisive criterion of truth. Nor do I doubt that people in a position to have emphatic experience have a moral obligation to speak on behalf of others. Social critique has long had a prophetic element, which philosophy would abandon at the cost of silencing itself. The problem in Adorno's vision, as I understand it, is that it makes philosophical experience self-authenticating.[22]

---

[22] Raymond Geuss gives drastic expression to this tendency in an essay titled "Suffering and Knowledge in Adorno," when he writes: "Nevertheless, in point of fact the standards of criticism have their concrete location in Adorno's sensibility. The society is said to fail because it fails him. . . . Adorno's philosophy can be seen as a philosophy of suffering spirit, a way of articulating the pain spirit experiences when confronted with a world that thwarts its aspirations, and as such, a criticism of that world." Although I think Geuss overstates his case and misses important nuances in Adorno's philosophy, I agree with him that Adorno needed to develop "a more sophisticated and differentiated analysis of suffering than he did."

His account of the "privilege of experience" makes philosophical experience speak for itself and entirely on its own authority. So no matter how much the articulation of such experience enters contexts of justification, the experience being articulated cannot really be challenged. In an odd and wholly unexpected way, Adorno seems to ground his philosophy in a negative version of the "authenticity" whose jargon he so effectively skewers.[23]

This problem directly affects his appeal to the need to express suffering. Recent ethnic conflicts and imperialist wars tell us that suffering does not speak for itself, and its expression is not self-authorizing. It is always already interpreted as the suffering of certain people in certain respects and with a certain measure of opprobrium or indignation. In 2001 North American media, for example, gave more weight to the sudden death of heroic fire fighters in New York City on 9/11 than to the decade-long starvation of hundreds of thousands of Iraqi children. The media's interpretation of suffering helped engender public support for a destructive invasion and occupation of Iraq that many would only later consider unjustified or even unjust. Hence, as I have said in a different context, "the need to express suffering cannot be self-evident as a condition of truth. The need must also be met in ways that are true." So too, the "philosophical recognition of this need, no matter how compelling, cannot be self-contained. The recognition must also represent those for whom the suffering is expressed and interpreted."[24]

Raymond Geuss, *Outside Ethics* (Princeton: Princeton University Press, 2005), pp. 115–16, 130.

[23] See especially Theodor W. Adorno, *The Jargon of Authenticity*, trans. Knut Tarnowski and Frederic Will (London: Routledge & Kegan Paul, 1973); *Jargon der Eigentlichkeit: Zur deutschen Ideologie* (1964), *Gesammelte Schriften* 6 (Frankfurt am Main: Suhrkamp, 1973), pp. 413–526. In the next chapter I compare and criticize Heidegger's conception of "authenticity" and Adorno's conception of "emphatic experience."

[24] Lambert Zuidervaart, *Adorno's Aesthetic Theory: The Redemption of Illusion* (Cambridge, Mass.: MIT Press, 1991), p. 306. Deborah Cook has asked whether my criticisms of Adorno ignore his insistence that suffering is objective, that suffering is "objectivity that weighs upon the subject" (*ND* 17–18/29). I do not deny that in suffering societal evil is inescapably registered. In that sense, suffering is objective. But I do question whether, in a philosophical context, suffering is immediate (unmediated) or its expression self-authorizing. If *Weh spricht: vergeh*, then there must always already be a language and an addressee for this expression, and one addressee will not automatically "hear" what someone else interprets as being said.

Accordingly, not even Adorno's formulation of a "new categorical imperative" speaks for itself. I realize how problematic it is to say this, and how questionable it is for me to say this. To say this seems immediately to violate the moment of ethical excess that Adorno's formulation is supposed to let one feel, thereby dishonoring the victims of Nazi genocide. Moreover, for me to say this seems to put me in a position of ethical superiority that I have neither the right nor the intention to claim. Yet Adorno himself acknowledges that truth claims made in the articulation of philosophical experience must enter contexts of justification if they really are to serve as truth claims. And his understandable reluctance to justify the claim that after Auschwitz everyone ought to prevent its recurrence does not keep him from giving an implicit justification. Implicitly, he argues that systematically inflicting physical pain on human beings is always abhorrent and that the societal conditions fostering and supporting such abhorrent conduct must be resisted and changed.

But Adorno casts this justification in the form of saying the new imperative, which he has formulated, lets one feel abhorrence despite the false consolations of post-Auschwitz culture. That manner of justification is problematic in two respects. First, it ignores the fact that abhorrence, although it is a corporeal feeling, is itself culturally informed and ethically inflected, such that psychopaths and sociopaths may seldom feel it. Second, his justification does not say why this feeling should have precedence over other feelings that also arise when people confront extreme suffering, such as anger, hatred, fear, despair, or compassion. Not even in circumstances of torture and cruelty does suffering speak for itself, at least not with respect to those who are not themselves the victims of torture and cruelty. The philosopher who takes seriously the need to express suffering also assumes an obligation to listen to other voices and to justify the philosopher's own expression with respect to those voices. The philosophical experience informing a social critique may be emphatic, but it cannot be self-authenticating.[25]

In other words, not even suffering can be removed from the dialectic of subject and object, which involves a further dialectic among subjects with respect to objects.

[25] In response to astute remarks made by Jon Short, I should clarify that I do not think Adorno makes *suffering as such* a self-authenticating event. He is not addressing all human suffering but that which arises from the societal logic of Western capitalism.

Objectification of Hope

The reverse side to Adorno's privileging of philosophical experience lies in his objectification of transformative hope. By "objectification" in this context I mean his tendency to regard hope as something whose basis lies in objects that have resisted the principles of identification and exchange. Hope comes to us, he seems to say, from a historically possible future in which objects would no longer be reduced to mere instruments and commodities, and from their potential even now to escape and resist the impositions of societally preformed, constitutive subjectivity. Further, such hope can arise because remnants in subjective experience remain as open to things in their nonidentity as does the child who delights in a favorite village as if it were completely unique (*ND* 373/366).[26]

Contrary to other critics such as Wellmer, my objection is not that Adorno heads toward an aesthetic emergency exit, that "he could ultimately only transfer the unthinkable thought of reconciliation to the realm of aesthetic experience" (*EG* 191/212). Although authentic artworks provide special sites for "the nonidentical" in this sense, I do not interpret Adorno as restricting such sites to art. This is clear even in the famous passage where Adorno links truth with hope and explicitly mentions art:

> Thought that does not capitulate before wretched existence comes to naught before its criteria, truth becomes untruth, philosophy becomes folly. And yet philosophy cannot give up, lest idiocy triumph in actualized unreason [*Widervernunft*].... Folly is truth in the shape that human beings must accept whenever, amid the untrue, they do not give up truth. Even at the highest peaks art is semblance; but art receives the semblance... from non-semblance [*vom Scheinlosen*].... No light falls on people and things in which transcendence would not appear [*widerschiene*]. Indelible in resistance to the fungible world of exchange is the resistance of the eye that does not want the world's colors to vanish. In semblance nonsemblance is promised. (*ND* 404–5/396–7)

Moreover, Adorno clearly acknowledges that such suffering needs to be *expressed*, and he makes its expression a condition of truth. The focus to my criticism lies in Adorno's tendency to let the *philosophical* experience of suffering speak for itself and to inoculate such experience against public discussion.

[26] Thanks to Jon Short for reminding me of this passage and insisting on the dialectical character of Adorno's emphasis on the objectivity of hope.

Adorno does not say that art is the only place where in semblance nonsemblance is promised, nor, as a materialist Hegelian, should he say this. The "priority of the object" implies that all objects in their nonidentity with the subject have priority, not simply those objects which draw artistic attention to their nonidentity. To hope for a future society without unfulfilled needs where suffering ends is to hope for more than art could ever deliver.

Instead, my objection is that objects as such are an inadequate basis for transformative hope, even in their dialectical relation with emphatic experience. Simon Jarvis has said that "the real possibility of reconciled non-identity is the speculative moment which animates each of Adorno's works" and "the condition of intelligibility of his very utterances and texts."[27] I think Jarvis is right. In my own terms, to eliminate hope for a future society without unfulfilled needs would be to remove the point of Adorno's negative dialectic as this culminates in his meditations on metaphysics after Auschwitz. But what makes "reconciled non-identity" a "real possibility"? What generates and sustains the hope for a society that is fundamentally different from this unspeakable world? Certainly Adorno considers thought's ability to think what lies beyond it in thinking against itself to be crucial in this regard. There must be something "more" than this, however, something about which thought can think, albeit in self-negation, and something that calls for such self-negating thought. Adorno's reply is that something *in der Sache* presses toward the concept of utopia (*ND* 404/396). Or, as his gloss on Kant's "intelligible world" suggests, something hidden to finite spirit turns itself toward spirit, which is compelled to think it (*ND* 392/384). This "something" is that in the object which is "not disfigured" (*ND* 57/66). It is that which, under current sociohistorical conditions, "does not exist and yet is not simply nonexistent" (*ND* 393/385). So, without claiming that this "more," this "something" actually exists, Adorno's negative dialectic in its speculative moment traces what in actual existence gives rise to the nonidentical's transient nonexistence.[28] And he counts on there being elements in human experience that do not "want the world's colors to vanish."

---

[27] Jarvis, *Adorno*, p. 230.
[28] See Deborah Cook, "From the Actual to the Possible: Nonidentity Thinking," *Constellations* 12 (2005): 21–35. Cook gives a Marxist reading of *Negative Dialectics* in

But this presupposes that somehow the nonidentical is there, *in der Sache*, historically and societally available to be traced by means of self-negating thought. It presupposes that something *in der Sache* really does press toward the concept of utopia. Quite apart from the epistemological questions this presupposition raises, it is a rather thin basis on which to hope for a fundamental transformation in society. Even if one transferred Adorno's presupposition to the side of the subject, suggesting, for example, that unmet material needs press for their satisfaction, it is not obvious why their satisfaction would not be endlessly deferred. Adorno's hope seems both crucial and ill-supported. The aporia in Adorno's metacritique of metaphysics is an internal conflict in the theme of hope and not simply an allegedly unsuccessful marriage between materialism and theology.

Totalized Transformation
Two tendencies in Adorno's negative dialectic give rise to this apparent aporia: first, his totalizing of transformation and, second, his failure to distinguish sufficiently between societal evil and the violation of societal principles. By "totalizing of transformation," I do not mean Adorno's central claims that contemporary society as a whole needs to be transformed and that this cannot occur unless the all-pervasive principle of exchange loses its grip. I share this position and think it is even more pertinent now, with the rapid globalization of capitalism, than it was when Adorno formulated it. Adorno's central claims mark a social-philosophical crossroad where Habermasian and non-Habermasian critical theorists part ways. I use the phrase "totalizing of transformation" to refer to Adorno's tendency to pit the transformation of society's entire architecture – society's deep structure, if you will – against transformations within that architecture – within social institutions, cultural practices, and interpersonal relations, for example. The tendency results in an all-or-nothing critique that, given the power of the exchange principle, makes "nothing" most likely for the foreseeable future. In other words, Adorno's radical critique of society is not radical enough.

order to take issue with the contrasting views of Adorno's "cognitive utopia" provided by J. M. Bernstein and Yvonne Sherratt.

It does not penetrate sufficiently to the multiple roots of change that together could generate the architectonic transformation he rightly envisions. To demonstrate this tendency would require a lengthier discussion than I can give here, and in any case other critics of Adorno have already made a similar point. But let me give one example of the problem I have in mind.

In the meditation on "Nihilism," Adorno claims that a theological consciousness of futility (*Nichtigkeit*) corrects those who believe that life here and now is meaningful, even if only in a few fulfilled moments. Yet he says the way to change the emptiness of life that theologians lament is not by people having a change of heart (*dass die Menschen anderen Sinnes werden*), but only by abolishing the life-denying principle, presumably, the principle of exchange. If that principle finally disappeared, so would the self-preservative cycle of fulfillment and appropriation (*ND* 378–9/371). I find Adorno's juxtaposition of personal conversion and structural transformation rhetorically clever but social-philosophically problematic.[29] How, pray tell, will the principle of exchange ever be abolished if the people who live under its grip and who sustain its operation do not have a change of heart? As Adorno has forcefully demonstrated elsewhere, we have long since left the stage of capitalism when, according to Marx and Engels, the proletariat had nothing to lose but its chains.

Earlier in *Negative Dialectics* Adorno himself suggests that fundamental transformation will require a "transparent solidarity" among human beings that currently is in short supply (*ND* 203–4/203–4). But if the architecture of society does not foster transparent solidarity, and if society's architecture needs to be changed out of such solidarity, then how will the requisite solidarity be fostered? Do not social institutions, cultural practices, and interpersonal relations all have a crucial role in this regard? Unless multiple roots of change can be found within the complex architecture of contemporary society – despite and by way of their entwinement with a capitalist economy – I see little prospect for abolishing the life-denying principle. Nor do I

---

[29] I discuss this problem under the heading "Antinomous Abstraction" in *Adorno's Aesthetic Theory*, pp. 85–8.

think such roots of change will flourish if the human participants do not undergo gradual and repeated "changes of heart."[30]

Adorno's inattention to plural sources of transformation within the differentiated fabric of contemporary society leads to his exaggerating the object in its nonidentity as a basis for transformative hope. Another way to put this is that Adorno has an insufficiently nuanced conception of "the subject." His conception oscillates between the societally constituted individual and the *Gesamtsubjekt* of not-yet-societally-actualized humanity. Lost in the oscillation are the diverse institutional, cultural, and interpersonal ways in which people become agents of change.[31]

Negative Utopia

The tendency to totalize transformation intersects a second source of Adorno's aporia, namely, his failure to distinguish sufficiently between societal evil and the violation of societal principles. By "societal principles" I mean historically developed, continually contested, and widely shared expectations about how social institutions should be organized, how cultural practices should be carried out, and how interpersonal relations should be configured. Justice, truth, and solidarity would be examples of such principles in contemporary Western societies. Human suffering can signal both societal evil and the violation of discrete societal principles. So returning to the need to express suffering will help uncover Adorno's tendency not to distinguish sufficiently between these.

I said earlier that the philosophical experience informing a social critique, no matter how emphatic, cannot be self-authenticating.

---

[30] Martin Morris, *Rethinking the Communicative Turn: Adorno, Habermas, and the Problem of Communicative Freedom* (Albany: State University of New York Press, 2001), pp. 158–91, recognizes the need for multiple roots of change and for personal transformation, but his notion of a "politics of the mimetic shudder" takes this recognition in an unduly restrictive, aesthetic direction. I have similar reservations about the notion of "fugitive ethical events" in J. M. Bernstein's *Adorno: Disenchantment and Ethics* – see my review of Bernstein's book in *Constellations* 10 (2003): 280–3. To the extent that a tendency to aestheticize resistance to societal evil pervades attempts at critically retrieving Adorno's negative dialectic, Wellmer's and Habermas's criticisms of Adorno's alleged aestheticism have a point. Nevertheless, I continue to find their criticisms misplaced. I return to the topic of society's fundamental transformation in Chapter 4.

[31] I return to the topic of collective agency in Chapter 6.

I also claimed that philosophers who take seriously the need to express suffering are obligated to hear and address other voices. Part of fulfilling this obligation, it seems to me, is to enter into conversation concerning the normative expectations according to which people recognize and respond to suffering. Adorno's own philosophy harbors the strong normative expectation that, in a world with abundant resources, no one's basic material needs should go unsatisfied. In itself, however, that expectation is hardly adequate either for identifying the full scope of what is wrong in contemporary society or for discovering "what needs to be done." Moreover, it would be the height of intellectual arrogance to provide a blueprint of how society should be changed, in order to satisfy basic material needs, without consulting those who have such needs. Nor could one consult the needy without attending to their own normative expectations concerning social institutions, cultural practices, and interpersonal relations.

Of course, Adorno is not about to provide any such blueprint. In that sense, the object of his hope is a "negative utopia,"[32] a society where no material needs are unmet, where needless suffering would end. Moreover, like Marx, Adorno sees all violations of societal principles, and perhaps even the principles themselves, as symptoms of a societal evil that resides in the structure of capitalist society as a whole. Precisely because he does not articulate distinct societal principles and does not explicate their specific violations, however, his object of hope becomes a displaced object – displaced into objects in their nonidentity. Few doors remain open to point people toward specific and shared expectations, articulable as societal principles, that would give substance to hope for the future. The scope of societal evil becomes so all-pervasive that discrete societal goods cannot be distinguished nor their particular absences thematized. Adorno's social philosophy leaves us empty-handed, hoping against hope that something *in der Sache* will not only continue to press toward the concept of utopia but also, in tandem with self-negating thought, enable that concept's actualization. This burden is more than any object or all objects could bear.[33]

---

[32] This is closely related to the description of Adorno's so-called utopianism as "utopian negativity" in Jarvis, *Adorno*, p. 222.

[33] In "The Possibility of a Disclosing Critique of Society: The *Dialectic of Enlightenment* in Light of Current Debates in Social Criticism," *Constellations* 7, no. 1 (2000): 116–27, Axel Honneth has tried to rescue Adorno and Horkheimer's mode of

The aporia of crucial but ill-founded hope arises from Adorno's inattention to multiple roots of fundamental transformation and from his failure to distinguish sufficiently between societal evil and the violation of discrete societal principles. Habermasian criticisms of Adorno try to correct these deficiencies, but at the price of removing the themes of suffering and hope from philosophy and becoming "postmetaphysical." To the extent that an aporia exists in Adorno's metacritique of metaphysics, it is, like its motivation, social-philosophical. Certainly the aporia has epistemological dimensions and implications, as Albrecht Wellmer has pointed out. But to address it would require more than an epistemological critique. It would require something like Adorno's effort to rescue within philosophy that which resists philosophy's own subsumptive concepts. It would require a social philosophy for which suffering is real and for which transformative hope is not misplaced. Perhaps such a social philosophy would show solidarity with Adorno's negative dialectic in the moment of its collapse.

> societal critique by construing their historico-philosophical framework as "a device of rhetorical condensation which a disclosing critique of society has to employ in order to evoke a new way of seeing the social world" (p. 124). Although Honneth is right to distinguish this manner of diagnosing social "pathology" from less global forms of normative social criticism, he avoids the question that troubled Adorno, namely, the basis for hoping that one day not only a *new way of seeing* but also a *new social world* will arise.

3

Heidegger and Adorno in Reverse

> A transformed philosophy ... would be nothing but full, unreduced experience in the medium of conceptual reflection.
>
> Adorno, *Negative Dialectics*[1]

The dialectical extremes of twentieth-century German philosophy touch in their conceptions of truth. More specifically, they touch in their conceptions of how truth is authenticated. Whereas Martin Heidegger says this occurs in the "authenticity" of Dasein, Theodor Adorno locates the authentication of truth in "emphatic experience." I wish to explore this dialectic, using Heidegger's *Being and Time* and Adorno's *Negative Dialectics* as my primary sources. Like Heidegger and Adorno, I consider truth to be a comprehensive idea that cannot be reduced to notions of propositional correctness or empirical accuracy. But I also regard correctness and accuracy as indispensable dimensions of truth. Similarly, I take authentication to be a comprehensive attestation of truth that cannot be reduced to discursive

---

An earlier version of this chapter was presented during the conference "Heidegger/Adorno: Aesthetics, Ethics, Technology" at the Université de Montréal in April 2004. I want to thank the conference organizers Iain Macdonald and Krzysztof Ziarek for their invitation and the participants for their instructive comments. Calvin Seerveld's comments on a first draft encouraged me to refine my formulations about the public authentication of truth. I also wish to acknowledge the collaborative support offered by the students in my graduate seminar "Truth and Authenticity: Heidegger's *Being and Time*."

[1] *ND* 13/25.

justification or verification. Yet justification and verification are inescapable ingredients of authentication.

My exploration has three stages. First, I summarize and criticize Heidegger's account of authenticity. Next, I examine Adorno's appeal to emphatic experience. Portraying "authenticity" and "emphatic experience" as each other's reverse image, I claim that neither one suffices to authenticate truth. Then, I draw out the significance of these dialectical extremes by proposing an alternative account of authentication.

3.1 EXISTENTIAL AUTHENTICITY

In *Being and Time*, Heidegger argues that the disclosedness (*Erschlossenheit*) of Dasein is the primary locus of truth.[2] This means that propositional truth, or the truth of assertions, is not the primary locus. Rather, whatever truth accrues to assertions and to the practice of making assertions stems from the fundamental openness that characterizes human relationships to other entities, to fellow human beings, and to oneself. Moreover, the empirical accuracy of our statements and claims stems from the "discoveredness" (*Entdecktheit*) that characterizes entities in relation to the disclosedness of the world Dasein inhabits. Accordingly, "only with the disclosedness of Dasein is the *most primordial* phenomenon of truth attained. ... In that Dasein essentially *is* its disclosedness, and, as disclosed, discloses and discovers, it is essentially 'true.' Dasein *is 'in the truth'*" (*SZ* 220–1).

For philosophers who maintain the primacy of propositional truth, Heidegger's emphasis on the disclosedness of Dasein has unsettling consequences. Ernst Tugendhat, for example, claims that Heidegger surrenders the concept of truth, even though he continues to use

---

[2] Passages of *SZ* in translation are taken from Martin Heidegger, *Being and Time*, trans. Joan Stambaugh (Albany: State University of New York Press, 1996). Page numbers refer to the pagination in *Sein und Zeit*, as found in the margins of the English translation. I have also consulted *Being and Time*, trans. John Macquarrie and Edward Robinson (New York: Harper & Row, 1962). I give preference to the Macquarrie and Robinson translation in retaining "Being" (capital "B") for "Sein" and in not hyphenating "Dasein." These modifications are made without comment in the citations and in my own text. Other relevant modifications to citations from the Stambaugh translation are marked by square brackets.

the word.³ This occurs because Heidegger equates truth and disclosedness without asking what distinguishes truth from untruth in various modes of disclosedness: "Heidegger has given the word truth another meaning. The broadening of the concept of truth, from truth as assertion to all disclosedness, becomes trivial if all that one sees in truth as assertion is the fact that it discloses in general."⁴ As a result, Tugendhat claims, Heidegger also surrenders the idea of critical consciousness. Yet Tugendhat recognizes the appeal of a conception "that, without denying the relativity and lack of transparency of our historical world, ... once again made possible an immediate and positive relation to truth: an alleged relation to truth that no longer stakes a claim to certainty, yet which also no longer poses a threat to uncertainty."⁵

There is something to Tugendhat's criticism. Yet, as I have tried to show in my book on *Artistic Truth*, Heidegger's conception has resources not only to counter this criticism but also to provide a more satisfactory account of propositional truth or assertoric correctness than the one Tugendhat assumes.⁶ Without rehearsing my arguments here, let me point out that Heidegger's discussion of truth has greater nuance than Tugendhat recognizes. At a minimum, Heidegger introduces five different ways in which truth can be distinguished from untruth.⁷ (1) The discoveredness of entities can be distinguished from their being covered up. (2) The disclosedness of the world and of Dasein can be distinguished from their lack of disclosedness. (3) The authenticity of Dasein's disclosedness can be distinguished from the inauthenticity of Dasein's disclosedness. (4) Dasein's falling prey within its disclosedness can be distinguished

---

³ Ernst Tugendhat, "Heideggers Idee von Wahrheit," in *Heidegger: Perspektiven zur Deutung seines Werks*, ed. Otto Pöggeler (Cologne: Kiepenheuer & Witsch, 1970), pp. 286–97; translated by Richard Wolin as "Heidegger's Idea of Truth," in *The Heidegger Controversy: A Critical Reader*, ed. Richard Wolin (New York: Columbia University Press, 1991), pp. 245–63. A longer version of this critique occurs in Ernst Tugendhat, *Der Wahrheitsbegriff bei Husserl und Heidegger*, 2d ed. (Berlin: Walter de Gruyter, 1970).
⁴ Tugendhat, "Heidegger's Idea," pp. 258–9; "Heideggers Idee," p. 294.
⁵ Tugendhat, "Heidegger's Idea," p. 261; "Heideggers Idee," p. 296.
⁶ See "Truth as Disclosure," in Lambert Zuidervaart, *Artistic Truth: Aesthetics, Discourse, and Imaginative Disclosure* (Cambridge: Cambridge University Press, 2004), pp. 77–100.
⁷ See *Being and Time*, §44, subsection b, especially SZ 220–3.

from Dasein's reclaiming itself from falling prey. (5) The illusion (*Schein*) and distortion (*Verstellung*) into which discovered entities sink (relatively to Dasein's falling prey) can be distinguished from their having been wrested from concealment. Provided such distinctions and their recognition need not have the same apparent rigor and certainty as the difference between the correctness and incorrectness of a simple assertion, Tugendhat's complaint is too crude. Whereas Tugendhat seems intent on deriving any broader conception of truth from an account of assertoric correctness, Heidegger aims to demonstrate the derivation of assertoric correctness from a more comprehensive ontology of truth.[8]

Authentic Disclosedness

Contra Tugendhat, Heidegger does not so much surrender the idea of critical consciousness as transpose it into the demand for authenticity. That is where the fulcrum to his ontology of truth lies. On the one hand, the discoveredness of entities and the disclosedness of Dasein are conditioned, at least in part, by the authenticity with which Dasein seizes upon Dasein's "potentiality-for-being-in-the-world" (*SZ* 228). On the other hand, Dasein can reclaim itself from falling prey and can wrest entities from illusion and distortion only to the extent that Dasein's own disclosedness is authentic. Heidegger signals the pivotal role of authenticity when he describes the possibility of authentic disclosedness: "This possibility means that Dasein discloses itself to itself in and as its ownmost potentiality-of-being. This *authentic* disclosedness shows the phenomenon of the most primordial truth in the mode of authenticity. The most primordial and authentic disclosedness in which Dasein can be as a potentiality-of-being is the *truth of existence*. Only in the context of an analysis of the authenticity of Dasein does it [the truth of existence] receive its existential, ontological definiteness" (*SZ* 221). The significance of the concept of authenticity is borne out by Heidegger's subsequent discussion of Dasein's "authentic potentiality-for-being-a-whole"

---

[8] According to Daniel O. Dahlstrom, *Heidegger's Concept of Truth* (Cambridge: Cambridge University Press, 2001), p. 392, Tugendhat tries to retain the "logical prejudice" that Heidegger's conception of truth aims to expose and dismantle.

(*das eigentliche Ganzseinkönnen*) in Division Two ("Dasein and Temporality," *SZ* §§45–83). There he states that his discussion results in a more complete grasp of that truth of Dasein which is "most primordial" because "it is *authentic*" (*SZ* 297).[9] Tugendhat, with his phenomenological notion of "evidence," gives too little attention to Heidegger's emphasis on authentication.

Without detailing Heidegger's elaborate "primordial existential interpretation" of Dasein in Division Two, let me briefly summarize his account of authenticity (*Eigentlichkeit*). His account calls attention to three topics: the ontological status of authenticity, the existential conditions of authenticity, and the primary characteristics of authentic disclosedness.

To begin with, Heidegger regards authenticity as ontological. The concept of authenticity pertains primarily to Dasein in its modes of existence, in its potentiality-of-being or ability-to-be (*Seinkönnen*), and not to actual attitudes, behaviors, accomplishments, or beliefs. Earlier, Heidegger had distinguished three equiprimordial modes of Dasein's existence: understanding, attunement, and talk. Of these three, understanding provides the primary (but not sole) locus of authenticity and inauthenticity. This implies that authenticity has a projective character and is future oriented. In its projective orientation to the future, authentic understanding aims at Dasein itself. Authenticity has to do with Dasein's understanding itself in terms of its ownmost (*eigenste*) possibility or potential rather than in terms of the world and others. When Dasein understands itself in this way, Dasein anticipates death. It anticipates death as that ownmost possibility which would render Dasein's existence impossible. This anticipated possibility is private, individualizing, unavoidable, certain, and indefinite. Authentic existence amounts to Dasein's being-its-self in an impassioned "freedom toward death."[10]

---

[9] As Dahlstrom (ibid., pp. 423–33) points out, the pragmatic readings of Heidegger proposed by Richard Rorty and Mark Okrent overlook the centrality of "authenticity" and "temporality" or "timeliness" (*Zeitlichkeit*) to Heidegger's conception of truth. This limits their usefulness as readings of Heidegger, even though they do provide an important counterweight to Tugendhat's criticisms of Heidegger's conception.

[10] It is worth quoting Heidegger's own summary of "authentic being-toward-death" in section 53 (*SZ* 260–7). The italics are Heidegger's: "*Anticipation reveals to Dasein its lostness in the they-self, and brings it face to face with the possibility to be itself, primarily*

This sheds light on the existential conditions that make authenticity possible. Because, to be authentic, Dasein must understand itself in terms of its ownmost possibility, nothing outside Dasein can make authentic existence possible. Rather, authentic existence is made possible by Dasein's own choice. It is made possible by Dasein's choosing to choose Dasein's ownmost potentiality-for-being-its-self rather than choosing to remain lost in the "they" (*SZ* 267–8). This potentiality-for-being-its-self is attested by conscience (*SZ* 279). In fact, such choosing to choose amounts to our wanting to have a conscience. Wanting to have a conscience is itself an understanding of Dasein's being directly called "to its ownmost potentiality-of-being-a-self" (*SZ* 269, 287–8). Moreover, the call of conscience is the call of care, a call Dasein gives to itself in its alienation from the public world (*SZ* 275–7). The self-given call of conscience summons Dasein to understand its own being-guilty as the null basis of its own potentiality-of-being (*SZ* 283–8). Choosing to choose Dasein's ownmost possibility, and hearing Dasein's own conscience, are the existential conditions of authenticity.

Accordingly, Heidegger characterizes authentic disclosedness (*Erschlossenheit*) as "resoluteness" (*Entschlossenheit*). He assigns three primary characteristics to Dasein's authentic potentiality-of-being (*eigentliches Seinkönnen*). These characteristics link back to the three equiprimordial modes of Dasein's disclosedness, to understanding, attunement, and talk. Authentic potentiality-of-being consists in (a) wanting to have a conscience (i.e., authentic self-understanding), (b) readiness for anxiety (*Angst*) as an attunement, and (c) reticence (*Verschwiegenheit*) or keeping silent (*Schweigen*) as a mode of talk. Taken together, these characteristic manners of understanding, attunement, and talk make up "resoluteness" as the "distinctive and authentic disclosedness" of Dasein (*SZ* 295–7).[11]

*unsupported by concern taking care of things, but to be itself in passionate anxious **freedom toward death** which is free of the illusions of the they, factical, and certain of itself*" (*SZ* 266).

[11] Heidegger summarizes the threefold existential structure of Dasein's authentic potentiality-of-being as follows: "The disclosedness of Dasein in wanting-to-have-a-conscience is thus constituted by the attunement of *Angst*, by understanding as projecting oneself upon one's ownmost being-guilty, and by [talk] as reticence. We shall call the eminent, authentic disclosedness attested in Dasein itself by its conscience – the *reticent projecting oneself upon one's ownmost being-guilty which is ready for* Angst – *resoluteness*" (*SZ* 296–7).

Perhaps we can say that resoluteness is what authenticates truth in Heidegger's conception. He puts it this way: "Now, in resoluteness the most primordial truth of Dasein has been reached, because it is *authentic*" (*SZ* 297). As the authenticating of Dasein's truth, resoluteness modifies the discoveredness of entities, the disclosedness of the world, and the concern of Dasein's being-with others (*SZ* 297–8). Resoluteness even "appropriates untruth authentically" (*SZ* 299). It reveals the authentic truth of existence. To this truth there corresponds an "equiprimordial being-certain" (*Gewisssein*) whereby Dasein unflinchingly and flexibly maintains itself in the actual factical situation disclosed by resoluteness (*SZ* 307–8).[12]

Such certainty has little to do with having the present under control or the past in our grasp. For resoluteness "is authentically and completely what it can be only as *anticipatory resoluteness*" (*SZ* 309). Moreover, authentication does not stop at the level of ontological structures. Just as anticipation (of death) is not simply an existential structure but an "existentiell potentiality-of-being" (*SZ* 309),[13] so too anticipatory resoluteness is not simply existential (i.e., an ontological structure) but also existentiell (i.e., an ontic way of life that a particular Dasein can embrace). As a way of life, anticipatory resoluteness disperses "every fugitive self-covering-over." It leads one to take action "without illusions," for it springs from a "sober understanding" of one's factical possibilities. "Together with the sober *Angst* that brings us before our individualized potentiality-of-being, goes an unshakable joy in this possibility" (*SZ* 310).

---

[12] Heidegger describes this "being-certain" as a "holding-for-true" (*Für-wahr-halten*) in which Dasein both gives itself to the situation and holds itself free for the possibility of taking itself back. In contrast with irresoluteness, "this holding-for-true, as a resolute holding oneself free for taking back, is the *authentic resoluteness to retrieve itself*." The ultimate certainty here is that resoluteness is constantly certain of death, which resoluteness anticipates. At the same time, anticipatory resoluteness gives Dasein "the primordial certainty of its being closed off," of its being constantly lost "in the irresoluteness of the they" (*SZ* 308).

[13] In calling this an "existentiell potentiality" (rather than simply a potentiality that is "existentielly attested"), I follow Macquarrie and Robinson's translation. Heidegger's German text reads: "das Vorlaufen ist ... der *Modus* eines im Dasein bezeugten existenziellen Seinkönnens." Macquarrie and Robinson (p. 357) translate: "anticipation is ... a *mode* of an existentiell potentiality-for-Being that is attested in Dasein." Stambaugh (p. 285) translates: "anticipation is ... a *mode* of a potentiality-of-being existentielly attested in Dasein."

Hence, the orientation with which Dasein inhabits its own disclosedness becomes decisive for Heidegger's conception of truth. Only the readiness and willingness and ability to face Dasein's own finitude and fallibility – not just once, and not simply upon occasion, but always again and anew – allows Dasein to be true and, in being true, to let other entities truly be. Dasein's truth can only be true insofar as it is authenticated.

Self-Denial

As I have said elsewhere, "there is something fundamentally right … about Heidegger's refusal to reduce truth to the correctness of assertions or the discoveredness of entities. He is correct not to exclude the ontological stance of those beings for whom truth itself, like Being, is a question and can never not be a question. Heidegger has successfully removed this question from the realms of Platonic perfection and Cartesian certainty. He has relocated it in those regions of human striving and disillusionment where getting things right often involves also getting them wrong, and where genuine discoveries seldom occur without difficult self-sacrifice."[14] Nevertheless, Heidegger's account of authenticity is problematic in three respects. First, it turns a substantial concept pertaining to actual merits into a formal state of being self-related. Second, it transfigures a historically conditioned and destructive rupture in the fabric of modern society (i.e., "alienation") into an ontological and authenticating encounter with one's own finitude. Third, it turns a mediated process of disclosure into a denial of mediation. Let me take up each problem in turn.

Formal Self-Relation

In ordinary usage, people describe something as authentic when it proves itself unique, or is particularly trustworthy (e.g., "the real thing"), or meets high expectations (e.g., "genuine"). To use the term in this way, one must already have sufficient dealings with the entity in question, both to detect its characteristic tendencies and to discriminate whether, in comparison with other entities or with other

---

[14] Zuidervaart, *Artistic Truth*, p. 96.

pathways open to this particular entity, its characteristic tendencies are particularly praiseworthy. Hence, whether one employs the term *eigentlich* in German or authentic in English, ordinary usage serves the making of substantial judgments about the merits of an entity or of its accomplishments.

Heidegger's account of authenticity exploits the nimbus of discrimination surrounding "authentic," as ordinarily used, to commend what is little more than a formal state of being self-related.[15] The formality of this state emerges in the ease with which Heidegger's account equates self-*understanding* with *deciding* ("choosing to choose"), *desiring* ("wanting to have a conscience"), and *adopting* modes of comportment (anticipation and resoluteness). Moreover, that which distinguishes such authenticity from inauthenticity is neither available for intersubjective judgment nor susceptible to "verification" by way of personal self-criticism. At best, the constituents of resoluteness (i.e., wanting to have a conscience, readiness for anxiety, and reticence) are predispositional states of consciousness or states of preconsciousness. As such they need have no intrinsic connection with the self's characteristic understandings, dispositions, and linguistic practices. Presumably one could want to have a conscience, be ready for anxiety, and be reticent in conversation without characteristically having a conscience, being anxious, or exercising conversational restraint. For the self to which one "relates" in resoluteness is little more than the possibility of a possibility – one relates to the possibility that one's own existence could be impossible. To call such a state of self-relation "authentic" is to forestall any assessments of the actual merits of that self and of its

---

[15] Adorno was especially allergic to the aura created by Heidegger's "authenticity" talk and the emptiness of what such talk commends. He objected that Heidegger turns the individual's decision to possess itself into the criterion of authenticity. This allows philosophy to ignore the real social conditions that make individuality possible, to avoid asking whether contemporary society actually allows people to be or become themselves, and to forget that "the old evil" (i.e., reification) might be concentrated in the Heideggerian concept of "selfness" (*Selbstheit*): "The societal relation that encapsulates itself in the subject's identity is de-societalized [by Heidegger's philosophy] into something in-itself." Theodor W. Adorno, *The Jargon of Authenticity*, trans. Knut Tarnowski and Frederic Will (London: Routledge & Kegan Paul, 1973), p. 115, translation modified; *Jargon der Eigentlichkeit: Zur deutschen Ideologie* (1964), *Gesammelte Schriften* 6 (Frankfurt am Main: Suhrkamp, 1973), pp. 489–90.

accomplishments, whether these assessments occur as self-criticism or as intersubjective judgment.

It is so, of course, that Heidegger does not intend his account of authenticity to provide criteria for self-criticism or intersubjective judgment. Yet his notions of self-understanding, anticipation, choosing to choose, and keeping silent make little sense apart from the notion of an individualized self that finds itself (and, according to Heidegger, reclaims itself) among other selves in a public world. Heidegger admits as much when he describes anticipatory resoluteness as not only *existential* but also *existentiell*.

But what does it mean for a self to find itself among other selves in a public world? Heidegger's own answer turns on the notion of letting others be themselves by wresting one's own self from falling prey to public talk. "[R]esoluteness toward itself first brings Dasein to the possibility of letting the others who are with it 'be' in their ownmost potentiality-of-being, and also discloses that potentiality in concern which leaps ahead and frees. ... It is from the authentic being a self of resoluteness that authentic being-with-one-another first arises, not from ambiguous and jealous stipulations and talkative fraternizing in the they and in what [the] they wants to undertake" (*SZ* 298). Accordingly, *inauthentic* existence is characterized by irresoluteness, by subservience to public understandings and interpretations, by "being at the mercy of the dominant interpretedness of the they. As the they-self, Dasein is 'lived' by the commonsense ambiguity of publicness in which no one resolves, but which has always already made its decision. Resoluteness means letting oneself be summoned out of one's lostness in the they" (*SZ* 299). For Heidegger, then, to find oneself among other selves in a public world is to remove oneself from that world and, in this removal, to endorse a similar removal on the part of others. But this amounts to *not* finding oneself among others as they are in public but rather finding others in relation to what one could be in one's own antipublic stance.

The problem, as I see it, is that Heidegger's formalism leaves little room for the self's authenticity to be either constituted or tested in public. Because the self's authenticity is neither constituted nor tested in public, it cannot authenticate truth in a public way. Insofar as the self's authenticity is Heidegger's primary and perhaps exclusive path for authentication, truth itself becomes a privilege of nonpublic existence. It is only a small step from this position to the even more problematic position that participation in an exclusive community is the proper path to authenticating truth.[16] To that extent, Adorno had good reason to attack the "jargon of authenticity" as a "German ideology."

Michael Zimmerman and other commentators have attempted to alleviate this problem by tracing it back to a "voluntarism" that Heidegger did not fully embrace in *Being and Time* and that he later abandoned when he transposed the notion of authenticity into that of "releasement" (*Gelassenheit*).[17] Such commentators might dismiss my criticisms as overemphasizing the voluntaristic elements in Heidegger's account of authenticity. But my criticisms do not revolve around the question of voluntarism. Even if Heideggerian authenticity were construed as a posture of acceptance, of "letting-be," rather than of resoluteness, of "choosing to choose," the self that either accepts or resolves would remain a nonpublic or antipublic self. And this undermines what I discuss later as the public authentication of truth.

Transfigured Alienation

Another way to say this is to claim that Heidegger's account of authenticity transfigures alienation. That is the second problem I mentioned earlier. The alienation of which Heidegger speaks occurs between Dasein and the public world. Discussing conscience as "the call of care," he writes: "[The caller] is Dasein in its uncanniness, primordially thrown being-in-the-world, as not-at-home, the naked 'that' in the nothingness of the world. The caller is unfamiliar to the

---

[16] Kevin Aho seems to miss the worrisome proximity of Heidegger's account of authenticity to Nazi ideology. Although I agree that Heidegger's account can be read in a more communalist and less individualist way, as Aho claims, I do not think such a reading removes the most problematic aspects of Heidegger's account. See Kevin Aho, "Why Heidegger Is Not an Existentialist: Interpreting Authenticity and Historicity in *Being and Time*," *Florida Philosophical Review* 3, no. 2 (Winter 2003): 5–22. In *Heidegger's Philosophy of Art* (Cambridge: Cambridge University Press, 2001), Julian Young suggests that by the mid-1930s Heidegger's site of authenticity shifts from individual Dasein to "great art," whose task is to secure an authentic "people" that actively appropriates its cultural heritage (pp. 52–60).

[17] Michael E. Zimmerman, *Eclipse of the Self: The Development of Heidegger's Concept of Authenticity*, rev. ed. (Athens: Ohio University Press, 1986). See especially the appendix, pp. 277–300.

everyday they-self, it is something like an *alien* voice. What could be more alien to the they, lost in the manifold 'world' of its heedfulness, than the self individualized to itself in uncanniness thrown into nothingness?" (SZ 276–7) At the core of authenticity, and voicing itself as Heideggerian conscience, is Dasein's alienation from its own everyday concerns, from the public world in which these concerns have their place, and from the public communications, perceptions, and interpretations that give these concerns their shape.[18] To be authentic, Dasein must be triply alienated, alienated from everyday concerns, from the public world, and from public interpretations of everyday concerns. For Heidegger, this "must" reflects not a historical condition but an ontological necessity.

Accordingly, when Heidegger elaborates his notion of conscience and contrasts this with other conceptions, he insists on a state of "being-guilty" prior to any responsibility or obligation. His is essentially an amoral conception of guilt. Dasein is guilty just by virtue of being Dasein and never gaining "power over one's ownmost being from the ground up" (SZ 284): "The summons [of conscience] calls back by calling forth: *forth* to the possibility of taking over in existence the thrown being that it is, *back* to thrownness in order to understand it as the null ground that it has to take up into existence. This calling-back in which conscience calls forth gives Dasein to understand that Dasein itself – as the null ground for its null project, standing in the possibility of its being – must bring itself back to itself from its lostness in the they, and this means that it is *guilty*" (SZ 287). Hence, the call of conscience is neither related to any specific deed nor critical with respect to specific courses of action.

Unfortunately this account turns a sociohistorical problem into an existential virtue. The very notion of an interior self whose authenticity resides in public withdrawal and perennial impotence is itself

---

[18] Compare in this connection Heidegger's description of Dasein's everydayness as involving idle talk, curiosity, and ambiguity (SZ §§35–38). Together these make up the "entanglement" of Dasein, its "character of being lost in the publicness of the they" (SZ 175). Shortly thereafter he uncovers anxiety as the attunement that throws Dasein back upon its own "authentic potentiality-for-being-in-the-world" (SZ 187). In anxiety one feels the "uncanniness" of "not-being-at-home" in the world. Anxiety individualizes by calling up Dasein's fundamental alienation from all that absorbs Dasein in its everydayness. See §40, SZ 184–91.

the philosophical expression of a modern cultural tendency whose societal matrix lies in the development of a market economy, privatized family life, and a depoliticized middle class.[19] It would not be difficult to find in Heidegger's characterization of authenticity the forms of alienation identified in Karl Marx's *Economic and Philosophical Manuscripts* and explained in Marx's subsequent writings. What Marx criticizes as societal ruptures – the alienation of workers from their labor, products, and fellow workers – Heidegger celebrates as ontological clues to the most primordial truth of Dasein. The result is the picture of a sober and anxious self whose inner authenticity renders it immune from both praise and blame for decisions and actions that are inescapably public. Not only does this self not find itself among others in public but it also secures itself against any challenge they might bring to the truth it claims and to the authentication it supposedly supplies.

For this reason I remain skeptical of attempts to find a basis for ethics in Heidegger's account of conscience. In an illuminating paper, Iain Macdonald argues that Heidegger's account offers us "an understanding of normativity rooted in non-identity."[20] The "nonidentity" lies in the difference between Dasein as it is and Dasein as it can be. Because of this difference, Dasein is always already "guilty."[21] Macdonald interprets such guilt to mean that individual Dasein exists in the gap between what it *is* and what it *ought to be*: "Dasein's self-identity contains an irreducible moment of negativity, of nonidentity, in the form of this gap between what Dasein is and what it can and ought to be. The gap is neither bridgeable nor fillable, and the difference it makes essential to existence always remains, no matter how Dasein pursues its projects." Because of this difference,

---

[19] See Charles Taylor, *Sources of the Self: The Making of the Modern Identity* (Cambridge, Mass.: Harvard University Press, 1989). Theodor W. Adorno's *Habilitationsschrift* on Kierkegaard, written partly in response to *Sein und Zeit* and published as a book in 1933, yields important sociocritical insights into Heidegger's emphasis on Dasein's *Innerlichkeit*. See *Kierkegaard: Construction of the Aesthetic*, trans. Robert Hullot-Kentor (Minneapolis: University of Minnesota Press, 1989); *Kierkegaard: Konstruktion des Ästhetischen, Gesammelte Schriften* 2 (Frankfurt am Main: Suhrkamp, 1979).

[20] Iain Macdonald, "Ethics and Authenticity: Conscience and Non-Identity in Heidegger and Adorno with a Glance at Hegel," in *Adorno and Heidegger: Philosophical Questions*, ed. Iain Macdonald and Krzysztof Ziarek (Stanford: Stanford University Press, 2007).

[21] Ibid.

Macdonald thinks Heidegger cannot be accused, à la Adorno, of "undialectically suppressing non-identity."[22]

I am not convinced by this attempt to rescue Heidegger, for three reasons. First, Heidegger's self-constituting "difference" does not lie between "is" and "ought" but between "is" and "can be." Nothing in his account of conscience, guilt, and resoluteness would turn a possibility into an obligation. Perhaps Heidegger thinks individual Dasein has an "obligation" to pursue what it can be rather than staying stuck in what it is. But such an "obligation" would be completely formal and open-ended, because my possibilities, while not infinite, are so varied that choosing any one of them, no matter how trivial and misguided, could count as my discharging the "obligation."

In the second place, to the extent that an "ought," an "obligation," surfaces in Heidegger's account, it is completely individual and self-related. Individual Dasein is called – indeed, calls itself – to be what it can be. Although this may seem to make me "responsible for saying what is or isn't right," as Macdonald puts it,[23] it does not make me responsible for *doing* either what *is* right or what *others* say is right. This is a strange sort of "responsibility," for it strips the self of any shared vocation, social ethics, and moral duty. Nor is it clear that an individual has a *responsibility* to pursue what "ought-to-be-for-me."

My third objection concerns the relation between guilt and responsibility. On Macdonald's construal, individual Dasein is responsible by virtue of being guilty, where guilt is understood as the unavoidable condition of the self's never being what it can be. Yet this gets matters precisely backward. If I am not already responsible for who I am and who I am becoming, then the gap between who I am and who I can be would not be a condition of guilt. It would simply be a gap between actuality and possibility, and deciding on one possibility over another would simply be an arbitrary choice. Nor would I be "answerable or accountable"[24] for such a choice, whether to myself or to others, unless I were already responsible, in some sense, for who I am and who I am becoming.

At bottom all three of my objections to Macdonald come to this: because Heidegger's nonidentity is a purely formal difference between the actual and the possible, it cannot be a source of normativity.

[22] Ibid.   [23] Ibid.   [24] Ibid.

In Kantian terms, to make individual Dasein the source of its only genuine norms is to reduce moral obligation to one's pursuing idiosyncratic maxims without asking whether those maxims are right.

Denial of Mediation

Yet it is precisely this alienated self, in its predispositional state of formal self-relation, that provides the authentication for truth in Heidegger's account. As I noted earlier, he regards the discoveredness of entities and the disclosedness of Dasein as conditioned by Dasein's authenticity. Only to the extent that Dasein's disclosedness is authentic can Dasein wrest entities from illusion and distortion and reclaim itself from falling prey. This means, however, that Heidegger's "most primordial" truth (*SZ* 221) is exactly the opposite of that complex mediation which his own critique of correspondence theories might lead one to expect. The most primordial truth is not a process in which various individuals and communities criticize, correct, and confirm each other's insights and dealings. Nor is it a process in which such insights and dealings are generated, tested, and revised by way of the entities to which they pertain. Rather, the most primordial truth is an anticipatory resoluteness whereby Dasein secures its own "freedom toward death" in disentanglement from the entities, including others, to which Dasein necessarily stands in relation. Hence, Heidegger's ontological key to truth is not mediation. Instead, it is Dasein's self-disentanglement from the mediations whereby it is itself constituted. And because the only orientation for such self-disentanglement is the possibility of Dasein's own death, the most primordial truth amounts not only to a denial of all that helps constitute the self but also to a self-denial.

Because of this displacement from mediation to disentanglement, Heidegger can consider Dasein to be "equiprimordially in truth and untruth" (*SZ* 223) and can claim that Dasein's resoluteness "appropriates untruth authentically" (*SZ* 299). In both of these formulations, untruth has to do with Dasein's unavoidable entanglement with entities and others that would divert Dasein from its "ownmost possibility." To think of these relations along the lines of entanglement and diversion, however, is to refuse those mediations without which, in my view, no self could be authentic and no truth could occur.

Lest this criticism of Heidegger seem hasty or unfair, let me quote at length from the pages that explain what the authentic appropriation of untruth comes to:

> As the they-self, Dasein is "lived" by the commonsense ambiguity of publicness in which no one resolves, but which has always already made its decision. Resoluteness means letting oneself be summoned out of one's lostness in the they. The irresoluteness of the they nevertheless remains in dominance, but it cannot attack resolute existence. . . . *For the they, however, [the] situation is essentially closed off.* The they knows only the "*general situation*," loses itself in the nearest "opportunities," and settles its Dasein by calculating the "accidents" which it fails to recognize, deems its own achievement and passes off as such. Resoluteness brings the being of the there to the existence of its situation. . . . [T]he call of conscience does not dangle an empty ideal of existence before us when it summons us to our potentiality-for-being, but *calls forth to the situation*. (SZ 299–300)

It might seem at first as if Heidegger here acknowledges the mediations with entities and others that help constitute the self. Yet it is only Dasein as a "they-self," not as a resolute and authentic self, that is so constituted: "Resoluteness means letting oneself be summoned out of one's lostness in the they." Again, it might seem as if such resoluteness is mediated with the public world on which resolutions depend. Yet the content of resoluteness does not arise from the situation upon which resolution seizes but rather from a conscience that calls one forth into the situation and allows Dasein to make its own "factical existence possible for itself" (*SZ* 300). The relation between Heidegger's resolute self and the public world is not one of mediation but rather one of self-disentanglement.

Hence I cannot fully endorse Taylor Carman's ingenious attempt to combine the two sides to Heidegger's account of selfhood. Taking a middle road between Guignon's "metaphysically optimistic" and Dreyfus's "pessimistic" construals of Heideggerian selfhood,[25] Carman reads Heidegger as both a "social externalist" (Carman's term) and an

---

[25] See especially Charles Guignon, "Philosophy and Authenticity: Heidegger's Search for a Ground for Philosophizing," in *Heidegger, Authenticity, and Modernity: Essays in Honor of Hubert L. Dreyfus, Volume 1*, ed. Mark A. Wrathall and Jeff Malpas (Cambridge, Mass.: MIT Press, 2000), pp. 79–101, and Hubert L. Dreyfus, *Being-in-the-World: A Commentary on Heidegger's* Being and Time, *Division I* (Cambridge, Mass.: MIT Press, 1991), pp. 299–340.

"ontological personalist" (my term). On the one hand, Heidegger's account of "the they" or "the one" (*das Man*) makes "anonymous social normativity" constitutive for all of Dasein's hermeneutic practices, including those of authentic self-interpretation. Although Dasein has "a structural tendency ... to lapse into banal, inauthentic interpretations of itself," established social practices have a "positive role" to play "in normatively structuring our practices and thereby constituting the intentionality of our everyday understanding."[26] Consequently, according to Carman's interpretation of Heidegger, "existing authentically does not consist simply in freeing oneself from all entanglements with the one, but rather in taking up a new, distinctive relation to the social norms always already governing one's concrete possibilities."[27] On the other hand, Heidegger's nonexpressivist and nonholistic account of authenticity emphasizes both the ontological irreducibility of a first-person perspective and "a profound asymmetry between first-person and second- and third-person modes of interpretation."[28] This first-person emphasis is so strong that Heidegger omits the social character of being a self and neglects the "other-oriented dimension of selfhood, indeed authentic selfhood."[29]

Although I agree with the main lines of Carman's double reading, I wonder whether he recognizes sufficiently how the gap between social externalism and ontological personalism vitiates Heidegger's account of authenticity. For the problem is not simply, as Carman puts it, that Heidegger's account remains silent about the hermeneutic conditions for bringing first- and other-person perspectives together "in an overarching interpretation of human beings as selves." The problem is not simply to say how it is possible "to come to understand myself, if only partially, as another" and thereby to engage "in empathy and imagination that is arguably essential to our mundane ethical self-understanding."[30] Instead, the problem is that Heidegger does not recognize – rather, he explicitly rejects – the constitutive role that others play in the emergence of a first-person perspective.

---

[26] Taylor Carman, *Heidegger's Analytic: Interpretation, Discourse, and Authenticity in Being and Time* (Cambridge: Cambridge University Press, 2003), p. 139.
[27] Ibid., p. 143.   [28] Ibid., p. 7.   [29] Ibid., p. 268.   [30] Ibid., pp. 268–71.

As Carman himself correctly observes, Heideggerian conscience "expresses and communicates ... an explicit recognition of the distinction between the everyday self of the one ... and the *proper* self, or one's *own* self, which is an ontological structure formally distinct from any of its own self-interpretations. ... Conscience calls Dasein away from all its ordinary self-interpretations back ... to the bare fact of its existence in all its concrete particularity."[31] But this presupposes, problematically, it seems to me, that there can *be* a "proper self" in its "concrete particularity" that is not always already, in its very "mineness," constituted by the relationships it sustains with other selves. It presupposes that one can have the perspective of "I" without always already having the perspectives of "myself" and "me," perspectives that do not emerge unless the "I" stands in relation to others and internalizes those relationships into its own "I-ness." From a Hegelian perspective, Heideggerian authenticity should be considered "abstract" rather than "concrete": social normativity must always remain *external* to this self and can never play a positive *normative* role in structuring the authentic self's existence. A Heideggerian authentic self can neither learn from social normativity nor contribute to it. This self is fundamentally asocial and therefore also ontologically impossible.

Let me summarize my three criticisms of Heidegger's account of authenticity. I have argued that his account reduces authenticity to a formal state of self-relation, transfigures historical ruptures in modern society into an ontological state of alienation, and turns the truth of Dasein into a denial of mediation. Because of the pivotal role "authenticity" plays in Heidegger's general conception, his idea of truth becomes internally untenable: despite the emphases on interdependence and intersubjectivity in his notions of "being-in-the-world" and "being-with," the most primordial truth of Dasein, whose own disclosedness is itself truth in the most primordial sense, lacks interdependence and intersubjectivity. Or rather, authenticity displays interdependence and intersubjectivity only in a privative way, as that from which Dasein must take distance in order to be authentic.

In other words, the possibilities opened up by Heidegger's expansive idea of truth as disclosedness get slammed shut by his

---

[31] Ibid., p. 293.

account of authentication. For if the authentication of truth depends upon Dasein's authenticity, and if Dasein can be authentic only in antipublic self-relation, then truth itself can no longer be attested in public. I would submit that, even on the most comprehensive conception of truth, what cannot be publicly authenticated is not truth at all.

3.2 EMPHATIC EXPERIENCE

It should be apparent that my criticisms of Heidegger resemble and in fact draw on Adorno's. Nevertheless, if one asks what account of authentication Adorno himself would offer instead of Heidegger's, one discovers a remarkable point of contact. This connection does not show up in the part of *Negative Dialectics* on Heidegger's ontology (*ND* 59–131/67–136).[32] Nor does it surface in the closely related ideology critique titled *The Jargon of Authenticity*. Rather the point of contact appears at the beginning and end of *Negative Dialectics*, in the introduction and the concluding "Meditations on Metaphysics," where Adorno takes on the issues to which Heidegger's account of authenticity responds. I shall concentrate on a passage from the introduction, already mentioned in the previous chapter, where Adorno explicates the concept of philosophical experience.[33]

---

[32] This part of *Negative Dialectics* is the weakest, in my view. It is susceptible to Brian O'Connor's comment, in his discussion of Adorno's critique of Heidegger, that Adorno sometimes misrepresents Heidegger "beyond recognition." Brian O'Connor, *Adorno's Negative Dialectic: Philosophy and the Possibility of Critical Rationality* (Cambridge, Mass.: MIT Press, 2004), p. 151.

[33] In the preface Adorno suggests that the entire introduction to *Negative Dialectics* "expounds the concept of philosophical experience" (*ND* xx/10). "Experience" is correctly identified as a "central concept of Adorno's thought" by the editor's introduction in *The Adorno Reader*, ed. Brian O'Connor (Oxford: Blackwell, 2000), pp. 1–19. In his subsequent book, O'Connor reads *Negative Dialectics* as "setting out to provide an account of experience that [Adorno] sees as exclusively expressible by a nonreified rationality." O'Connor, *Adorno's Negative Dialectic*, p. 13. O'Connor interprets the epigraph to the current chapter – concerning "full, unreduced experience in the medium of conceptual reflection" – as referring to experience, rather than to what Adorno explicitly identifies as "a transformed philosophy" (*eine veränderte Philosophie*). This misinterpretation leads to an account that, unlike Adorno's, renders experience overly conceptual, it seems to me. See, for example, O'Connor, *Adorno's Negative Dialectic*, pp. 3 and 79.

Negative Dialectic

In general, Adorno's negative dialectic is a philosophical attempt to conceptualize the nonconceptual without subsuming the nonconceptual under a system of concepts. As such, negative dialectical philosophy must rely on experience that provides access to the nonconceptual and is neither conceptually prescribed nor incompatible with conceptuality. The experience on which philosophy must rely can be called "emphatic experience." Following J. M. Bernstein, I use the term "emphatic experience" (*Erfahrung*) to refer to something the subject undergoes in relation to a particular object in its nonconceptualized particularity. Emphatic experience is characterized by novelty and by the object's directing the subject's response. It involves "a transformation of the individual [subject] and the emergence of a new object domain."[34] Adorno sees modern societies as having diminished the possibility and authority of such experience. That is why what he calls "philosophical experience" is crucial. Philosophical experience would reconnect rationality with emphatic experience while making palpable its modern demise. Adorno's concept of "philosophical experience" points to the complex ways in which the mediation of conceptualization and the nonconceptual occurs.[35]

Inevitably the question arises whether Adorno's appeal to experience is arbitrary. He takes up this question in the section titled "Privilege of Experience" (*ND* 40–2/50–3). The section begins by claiming that "the objectivity of dialectical knowledge" requires more from the epistemic subject rather than less. The standard "positivist" objection to this requirement would be that the requirement is "elitist and undemocratic." It is elitist and undemocratic, says the objector,

---

[34] J. M. Bernstein, *Adorno: Disenchantment and Ethics* (Cambridge: Cambridge University Press, 2001), p. 115. For a discussion of the multifaceted character of "experience" in Adorno's writings, see Martin Jay, "Is Experience Still in Crisis? Reflections on a Frankfurt School Lament," *Kriterion*, no. 100 (December 1999): 9–25. A corrected version of this essay appears in *The Cambridge Companion to Adorno*, ed. Tom Huhn (Cambridge: Cambridge University Press, 2004), pp. 129–47, and is incorporated into the discussion of Adorno in Martin Jay, *Songs of Experience: Modern American and European Variations on a Universal Theme* (Berkeley: University of California Press, 2005), 343–60.

[35] The need to reconnect rationality with emphatic experience is also central to Adorno's insistence on the possibility and necessity of what he calls "metaphysical experience," which I discuss in Chapter 2. See also Bernstein, *Adorno*, pp. 415–56.

because the "experience" on which dialectical objectivity supposedly depends is itself the prerogative of individuals who have a particular ability and biography. How can "philosophical experience" be a condition of knowledge if not everyone is capable of such experience?

Adorno immediately concedes that not everyone is capable of philosophical experience. But he says the standard objection ignores a fundamental reason for this incapacity, namely, how "the administered world" intellectually cripples the inhabitants of its iron cage. Only those who resist the pressure to fit in can challenge such a society, and their number is limited. In fact, the "privilege" of having experiences from which to critique an undemocratic society is created by that very society: "Critique of privilege has become a privilege: that is how dialectical the course of the world is" (*ND* 41/51). Conversely, to expect that everyone in this society could understand everything worth noticing would be to orient knowledge to "pathological traits" of people whose capacity to have experience (*Erfahrungen zu machen*) has been destroyed "by the law of perpetual identity [*Immergleichheit*]" (*ND* 41/51). Although seemingly democratic, this expectation of public intelligibility would actually undermine the critique needed to counter antidemocratic tendencies.

The privilege of experience is not really elitist, Adorno suggests, because it comes with a moral obligation. Those who, despite the constraints of prevailing norms, are capable of experience that gives rise to social critique have a moral obligation to express what most people cannot see. Expressing this is a "representative" (*stellvertretend*) effort in which some speak on behalf of many. Nor would "elitist pride" befit philosophical experience, because the capacity for such experience is largely an accident of history and results from the way class-based society is structured (*ND* 42/52).

Still, one could ask whether philosophical experience suffices. Should not the critique to which philosophical experience gives rise be subject to public discussion? And what about the experience itself? Should not it be subject to intersubjective testing? Adorno suggests two responses to this line of questioning. The first is simply to say, Do not confuse truth with public intelligibility: "The criterion of what is true is not its direct communicability to everyone. ... Truth is objective and not plausible" (*ND* 41/51–2). Adorno's second response is that truth mediated through experience loses its supposedly privileged character by becoming publicly discussable. This occurs when philosophical truth claims do not make special pleas for the experiences that give rise to them, but instead enter "configurations and contexts of justification [*Begründungszusammenhänge*]" that either bear them out or establish their inadequacies (*ND* 42/52). So philosophical experience gives rise to claims whose truth can be debated, even though what makes them true is not their public intelligibility.

Yet Adorno glosses "philosophical experience" in a way that raises the worry of elitism he has just tried to allay. "Within philosophical experience chances that the universal randomly grants individuals turn against the universal that sabotages the universality of such experience. Were this universality achieved [*hergestellt*], the experience of all individuals would change accordingly and would lose much of the contingency that meanwhile continues fatally to disfigure their experience" (*ND* 42/52). In other words, thanks to the way society is structured, certain individuals are capable of an experience that challenges this structure and its distorting of other people's experience. To the extent that other people's experience is distorted, not accidentally, but necessarily by virtue of society's structure, it would appear that the experience that challenges this structure is itself immune from intersubjective testing. Would not intersubjective testing itself be suspect, Adorno seems to ask, insofar as this occurs under distorting societal conditions?

Unintelligible Truth

This looks to me like a reverse image to Heidegger's account of authenticity, including its attendant problems. Like authenticity in Heidegger's ontology, emphatic experience is supposed to authenticate truth in Adorno's negative dialectic. But just as Heidegger's concept of authenticity renders truth immune from public authentication, so Adorno's idea of truth cordons truth-authenticating experience off from intersubjective testing. Each of the three problems already noted in Heidegger's account of authenticity has a reverse image in Adorno's account of philosophical experience. Whereas Heidegger turns a substantial concept into a formal state of self-relation, Adorno derives a substantial experience of the nonconceptual from a formal concept of societal structure. Whereas

Heidegger transfigures sociohistorical ruptures into an ontological structure, Adorno transforms particular experiences of such ruptures into universal sources of critical insight. Whereas Heidegger turns a mediated process of disclosure into a denial of the self's mediation, Adorno turns an affirmation of the self's mediation into a restriction on the process of disclosure. Let me indicate how each of these problematic reverse images shows up in Adorno's text.

Universal Abstraction
The first problem in Adorno's account of philosophical experience pertains to the relation between societal structure and substantial experience. Adorno regards the structure of late capitalism as highly abstract and as one that operates through universal abstraction. In permeating everyday life, such abstraction "sabotages the universality" of experience. Yet at the same time this very same structure "randomly grants individuals" chances that "turn against" universal abstraction. And from this random bestowal an experience somehow emerges that resists universal abstraction. Adorno considers this experience sufficiently substantial to secure access to the nonconceptual that must be conceptualized but not systematized.

I find Adorno's own account of this relation highly abstract, quite unlike the very specific comments he makes about the ideological positioning of Heidegger's "jargon of authenticity." Adorno refrains from identifying the sociological roles and cultural traits that would be prerequisites for having philosophical experience. Although he waves his hand at the class structure that remains in effect, he makes no effort to acknowledge the differential insights various class positions could afford. The transition from universal societal structure to particular experience has a nearly magical quality. His claim that the experience of certain unnamed individuals is an exception, an exception made possible by the universal rule of abstraction, threatens to become a mere assertion.

Self-Authentication
Nevertheless, Adorno repeatedly claims that philosophical experience, of which many people have been made incapable, is a crucial source of critical insight into the society that incapacitates them. But more than a crucial source, it is also a universal source, in two

respects. First, it puts the one who has philosophical experience in a position to speak on behalf of all others. It entitles – indeed, morally obligates – the critical theorist to make this "representative" effort. Second, philosophical experience comes with the presumption that the insight it generates is universally true. In fact, there would likely be no moral obligation to express what most people cannot see if philosophical experience were not presumed to generate universally true insight.

Unfortunately, to turn particular experiences of sociohistorical ruptures into universal sources of critical insight is to cordon these experiences off from intersubjective testing. The problem, as I explained in the previous chapter, is that Adorno has made philosophical experience self-authenticating such that the experience being articulated cannot be challenged. The "elitist" element in Adorno's account of philosophical experience is not that it puts some people in a position to speak on behalf of everyone else. The elitist element is that the representative is both self-nominated and self-elected. This is not simply a matter of careless formulation on Adorno's part. By describing others as incapacitated by the societal system, he has effectively disqualified them as participants in the critical process. Or, rather, he has acceded to the disqualification purportedly carried out by society itself. Such disqualification comes back to haunt the position that declares others disqualified. For if *they* are disqualified, then the "representative" effort also loses its point, and the quality of the critical theorist's experience no longer matters. Experience that is self-authenticating in this way cannot be a universal source of critical insight.

Esoteric Index
Compared with Heidegger's account of authenticity, Adorno's account of philosophical experience has the advantage that it affirms rather than denies the objective mediation of the self. Adorno rightly criticizes Heidegger's account for ignoring such mediation. According to Adorno, it is because society mediates the self that truth requires subjective mediation. When he develops this emphasis on the self's mediation in his account of philosophical experience, however, he does so at the expense of truth's public intelligibility. A dramatic instance occurs when Adorno declares: "Today every step toward communication sells out the truth and makes it false" (*ND* 41/51–2).

One wonders how this rather jaundiced view of public intelligibility jibes with Adorno's subsequent acknowledgment that truth claims arising from philosophical experience need to be justified. Presumably the "contexts of justification" are ones where public intelligibility is required. Yet Adorno makes it seem that the public intelligibility of a philosophical truth claim has no direct bearing on the truth of the claim or on the truth of the experience from which the claim arises.

To the extent that this is Adorno's position, his idea of truth becomes untenable. It is made untenable by his regarding self-authenticating experience as the way to authenticate truth. Adorno tries to forestall this consequence by saying that the interwoven texture (*Geflecht*) of truth is "the index of itself" (*ND* 42/52). In saying this he subscribes to a holistic idea of truth, as Heidegger does with his insistence on disclosedness. But just as Heidegger undermines his ontological idea of truth by seeking authentication in antipublic resoluteness, so Adorno weakens his negative dialectical idea of truth by seeking authentication in publicly unintelligible experience. For if the strands that supposedly make up the texture of truth cannot be contributed or checked by those on behalf of whom the dialectical thinker weaves or unfolds this texture, then truth will be an index that few will consult. We shall be left with what Habermas and Wellmer describe as an esoteric idea of truth.

The problem with an esoteric idea of truth is not that it is "elitist and undemocratic," charges that people too easily trot out when they understand a critique but wish to ignore it. Rather, the problem is that an esoteric idea of truth cannot do justice to truth itself as a mediated process in which everyone has a stake and outside which no one in contemporary society can flourish. Nor can truth be authenticated by publicly unintelligible experience that an abstract societal structure randomly grants, even though this experience claims to be a universal source of critical insight. While correcting Heidegger's account of authenticity, Adorno shares his failure to provide for the public authentication of truth.

3.3 PUBLIC AUTHENTICATION

Earlier I distinguished truth from the authentication of truth. But that distinction is not transparent. Although one can speak of truth that has not been authenticated, it is difficult to imagine authentication that does not already claim to be true. So the distinction between truth and authentication cannot assign equal weight to each side. Rather, authentication must be regarded as an extension of truth. Although the unfolding of truth relies on this extension, the direction of truth's unfolding does not derive from authentication. Hence, to work out a viable conception of authentication, one must begin with a sufficiently comprehensive idea of truth.

Elsewhere I have characterized truth as an indissoluble and dynamic correlation between human fidelity and societal disclosure.[36] I have qualified the disclosure of society as "life-giving disclosure." By this I mean a societal process in which human beings and other creatures come to flourish in their interconnections. I have also specified human fidelity as faithfulness to societal principles such as justice and solidarity that commonly hold for people and that people hold in common. The correlation between disclosure and fidelity is indissoluble: creaturely flourishing depends in part on how human beings pursue fidelity to the commonly holding/held, and the telos of their fidelity is to promote the interconnected flourishing of human beings and other creatures. The correlation is also dynamic: it is a continually unfolding process, not a static structure. So too, the principles in question are historical horizons, not timeless absolutes: they emerge from social struggles in which such principles are always already at stake.

Regarded in this way, truth can be authenticated if cultural practices and social institutions enable people to bear witness to a correlation between societal disclosure and their fidelity to societal principles. But what people attest cannot simply be the general process of correlation. It must be specific correlations that occur as people engage in cultural practices and participate in social institutions. In that sense, as Heidegger claimed about authenticity, authentication is both existential and existentiell. Because authentication is supported by cultural practices and social institutions, however, and because it occurs through involvement with these practices and institutions, authentication is intrinsically intersubjective, unlike Heideggerian authenticity. At the same time, one

---

[36] Zuidervaart, *Artistic Truth*, pp. 96–100.

must recognize that society as a whole can be so distorted that specific correlations between societal disclosure and principial fidelity are the exception rather than the rule. In that sense, Adorno's appeal to "emphatic experience" is not misplaced. Yet, because the possibility and occurrence of these correlations depend upon practices and institutions sustained by that same society, the "exceptions" cannot be attributed to a "privilege of experience." They are made possible by the very principles and disclosive process that a distorted society occludes. In other words, authentication depends on truth. As both Heidegger and Adorno acknowledge, truth cannot be exceptional.

What does it mean to bear witness to correlations between disclosure and fidelity? It does not mean simply to articulate concerns in language and to raise and defend certain validity claims, even though linguistic and discursive practices are indispensable ingredients. Rather, to bear witness is to participate in such correlations in a manner that invites both oneself and others to do the same. If, for example, the correlation of contemporary justice and human flourishing requires the elimination of systemic racism, one bears witness to this correlation by doing what one can, with others, to transform the racist practices and institutions to which one belongs, whether through gestures, policies, or public protests. To bear witness to the truth means to do what truth requires in a social context and with respect to others who co-inhabit that context. Bearing witness involves the full range of human activities, not only linguistic and discursive but also aesthetic, ethical, political, economic, and the like.

Whereas truth is the processual correlation between fidelity to principles and life-giving disclosure, the authentication of truth is an invitational enactment of specific correlations in particular circumstances. It is the invitational quality of this enactment that makes authentication public and not privileged. For an invitation always welcomes a response from those invited, including the one who invites. And to be an open invitation rather than a demand or statement or idiosyncratic gesture, it must invite uncoerced acceptance or rejection or inattention.

Describing authentication in this manner casts a different hue on those modes of authentication which Western philosophy has valorized. I think here of a standard emphasis on discursive justification and verification. Earlier I claimed that justification and verification are inescapable elements of authentication to which authentication cannot be reduced. Now I wish to add that, as elements of authentication, justification and verification themselves are at bottom invitational enactments of correlations in context. What distinguishes them from other modes of authentication is their peculiar structure, which singles out the purported universality and necessity of enacted correlations. Such discursive practices try to bear out the purported universality of the validity claims raised in other linguistic practices. They try to do this with an appeal to their own inherent validity. Their inherent validity is not a matter of other societal principles such as solidarity and justice but rather a matter of logic and rhetoric. Hence, the appearance arises that justification and verification are not modes of authentication and that what gets justified or verified has little to do with the comprehensive idea of truth. Moreover, given a pervasive "logical prejudice" (Dahlstrom) in Western philosophy, which both Heidegger and Adorno challenge, it also appears that the comprehensive idea of truth has little to do with truth "properly speaking" – that is, with truth as correctness or accuracy.

Such appearances turn truth upside down. What drives discursive practices and makes them important is their role within multidimensional processes of authentication. Part of invitationally enacting correlations of fidelity and disclosure is to test the reach of the principles at stake and to establish the extent of the circumstances in which correlations are enacted. It makes a difference for the enactment itself whether justice is only for "just us" or also for "others," whether the flourishing of some comes at the expense of others. It also matters that we have discursive practices and discursively attuned institutions within which deliberation about such differences can occur. To accomplish this, neither prelinguistic intuitions nor postdiscursive decisions suffice.

Yet it would be a mistake to think that discursive practices have the final word or that they transcend the contexts of authentication. Discursive practices occur within multidimensional processes of authentication; other modes of authentication provide their context. This implies that conflicts can occur between discursive and nondiscursive modes of authentication and that such conflicts cannot always be resolved in a discursive fashion. What is logically true can be contextually false: a "good argument" can support unjust and destructive arrangements. Conversely, what is "true in practice" may receive articulations and defenses that are logically invalid.

Nevertheless, the normative role of justification and verification as modes of authentication is to test the universality of societal principles and the necessity of specific correlations between principial fidelity and societal disclosure. Is the contemporary principle of justice such that systemic racism must be eliminated? And are particular gestures, policies, or protests necessary in that regard? Although one does not need well-articulated answers to such questions in order to challenge systemic racism, resistance would falter if the questions never arose.

The structurally peculiar focus on validity, combined with Western philosophy's logical prejudice, has led many philosophers to link the public character of authentication with the "rationality" of discursive practices. Both Heidegger and Adorno question that linkage. They do so, however, at the expense of authentication's public character. I have proposed, in a preliminary fashion, to retain their concern for the comprehensive authentication of truth without either surrendering the public character of authentication or reducing authentication to discursive practices.[37] Truth as such may not be democratic, but its invitational enactment must be public. A public invitation will be open to free recognition and acceptance or refusal on the part of those invited. To that extent, and to the extent that public freedom, recognition, and participation are the hallmarks of democracy, the authentication of truth must be not only public but democratic. Truth calls for public authentication. It calls for democratic truth telling, both verbal and nonverbal, that does not avoid public presentation and response.

It is precisely because Heidegger and Adorno emphasize nondiscursive authentication that their contributions are important and their hidden point of contact deserves further attention. Both

---

[37] This discussion of public authentication implies a refusal to privilege discourse in the manner of Habermas's discourse ethics and, albeit less prominently, of his more recent pragmatic theory of truth. See Jürgen Habermas, *Moral Consciousness and Communicative Action*, trans. Christian Lenhardt and Shierry Weber Nicholsen, introd. Thomas McCarthy (Cambridge, Mass.: MIT Press, 1990), and *Truth and Justification*, ed. and trans. Barbara Fultner (Cambridge, Mass.: MIT Press, 2003). I hope to discuss Habermas's theory of truth in a subsequent publication.

Heidegger and Adorno recognize that philosophical conceptions of truth have far-reaching implications for the kind of society we inhabit and the sorts of people we become. They make compelling cases for the claim that traditional Western conceptions of truth, whether classical and metaphysical or modern and epistemological, have supported destructive tendencies in society, while also opening up potentially fruitful paths. Neither one thinks "truth" can be left to the logicians and the technicians of philosophy. Each has attempted in his own way to reconnect the idea of truth with the cultural issues and social crises from which overly professionalized and hyperspecialized philosophies take distance. Their assessments may be dramatically opposed, but together Heidegger and Adorno have placed such matters at the center of philosophical attention.

The concern Heidegger and Adorno share for the authentication of truth needs to be understood in this societal context. Both of them recognize that truth is not simply a theoretical concern, that truth must be borne out in contemporary lives and practices and institutions. Both of them claim that the historical trajectory of modern Western society makes this requirement increasingly difficult to sustain. Both of them seek a site from which that trajectory can be resisted and perhaps redirected, whether the site be the authenticity of Dasein or the occurrence of emphatic experience.

But these sites are oppositional rather than transformative, and their access is restricted rather than open to a broader public. As I have tried to suggest, oppositional and restricted sites of authentication are inadequate for the comprehensive process of truth that both Heidegger and Adorno endorse. Their philosophies leave us with an exceptional challenge: to discover how truth can be borne out in ways that are authentic, emphatic, and thoroughly democratic. In such modes of public authentication, and in a society that sustains them, the dialectical extremes of Heidegger and Adorno would not only touch but be true. No longer restricting the ranges of experience that can authenticate the truth of its conceptual reflection, philosophy itself would be transformed.

4

Globalizing Dialectic of Enlightenment

> Enlightenment ... has always aimed at liberating human beings from fear and installing them as masters. Yet the wholly enlightened earth radiates under the sign of disaster triumphant.
> Horkheimer and Adorno, *Dialectic of Enlightenment*[1]

Dialectic of Enlightenment deserves the label Adorno gave the *Missa Solemnis*: "alienated masterpiece." With only one modification, his comments on Beethoven's celebrated and misunderstood composition can express the reception accorded Adorno and Horkheimer's work: "Every now and then ... it is possible to name a work in which the neutralization of culture has expressed itself most strikingly; a work ... which occupies an uncontested place in the repertoire even while it remains enigmatically incomprehensible; and one which ...

Earlier versions of this chapter were presented on three occasions: as a conference paper for "Secularity and Globalization: What Comes after Modernity?" at Calvin College in November 2005; as an invited lecture sponsored by the Department of Philosophy at the University of South Carolina in the same month; and as a keynote address to the Canadian Society for Continental Philosophy during the annual Congress of the Humanities and Social Sciences in May 2006. I want to thank James K. A. Smith, Otávio Bueno and Martin Donougho, and Diane Enns and Paul Fairfield, respectively, for arranging these occasions, as well as the audiences for their provocative questions. In particular I wish to acknowledge generative comments made by Ron Kuipers, Jerry Wallulis, and Graham Ward.

[1] *DE* 1/25; translation modified. My discussion in this chapter and the next will focus on Adorno's ideas, even though he and Horkheimer cowrote substantial portions of the book.

offers no justification for the [abuse] accorded it."² As Adorno observed with respect to the *Missa Solemnis*, "to speak seriously" of *Dialectic of Enlightenment* today "can mean nothing less than ... to alienate it." Here, however, alienating the work does not require one "to break through the aura of irrelevant worship which protectively surrounds it,"³ but to disrupt gestures of easy dismissal that prohibit a thoughtful engagement, at a time when debates about globalization increase its significance.

4.1 HABERMAS'S PARADIGMATIC CRITIQUE

It is not so much Adorno's opponents as his own successors who make a fresh reception difficult. There is by now a veritable army of critical theorists who regard *Dialectic of Enlightenment* as the nadir of the Frankfurt School from whose abyssal aporias all must be rescued. The marching tune of this protective phalanx is familiar. It begins with Jürgen Habermas's muted worries in the late 1960s about an alleged attraction to a mystical "resurrection of fallen nature." As Habermas put it in a public tribute shortly after Adorno's death, "Adorno ... entertained doubts that the emancipation of humanity is possible without the resurrection of nature. Could humans talk with one another without anxiety and repression unless at the same time they interacted with the nature around them as they would with brothers and sisters?" Over against Adorno as a latter-day Saint Francis, Habermas's own stance is firm: "The concept of a categorically different science and technology is as empty as the idea of a universal reconciliation is without basis."⁴ Citing Albrecht Wellmer, Habermas

---

² Theodor W. Adorno, "Alienated Masterpiece: The *Missa Solemnis* (1959)," *Telos*, no. 28 (1976): 113–24; quotation from p. 113, with the word "abuse" (in square brackets) substituted for "admiration."

³ Ibid., p. 113.

⁴ Jürgen Habermas, "Theodor Adorno: The Primal History of Subjectivity – Self-Affirmation Gone Wild (1969)," in *PP* 101–11/184–99; quotation from p. 109/196. For similar passages in other publications from the 1960s, see *Knowledge and Human Interests* (1968), trans. Jeremy J. Shapiro (Boston: Beacon Press, 1971), pp. 32–3, and *Toward a Rational Society: Student Protest, Science, and Politics* (1968, 1969), trans. Jeremy J. Shapiro (Boston: Beacon Press, 1970), pp. 85–6. It is interesting that already in 1963, in a tribute to Adorno on his sixtieth birthday (not included in the English translation of *Philosophical-Political Profiles*), Habermas portrays Adorno's intellectual biography and *Dialectic of Enlightenment* as centrally ambivalent about

links Adorno's ambivalent hope for reconciliation with a tendency to turn Marx's critique of political economy into a philosophy of history that neglects an empirically based theory of late capitalist society.[5]

From these opening strains, the battle cry swells into increasingly hostile and hermeneutically astonishing attacks. Albrecht Wellmer, in an otherwise careful and instructive essay, sees in *Dialectic of Enlightenment* an "anthropological and epistemological 'monism'" whose single-minded rejection of "instrumental reason" as "the paradigm of perverted reason" threatens to turn liberation into "an eschatological category."[6] Around the same time, Axel Honneth portrays Adorno's "critical social philosophy" as an exercise in "pessimistic self-clarification" that "cannot commit itself to an idea of historical progress which goes beyond total reification."[7] Relying heavily on the interpretations of Grenz[8] and of Baumeister and Kulenkampff,[9] Honneth claims that Adorno sees "aesthetic theory and negative dialectics" as "the only means whereby a weakened critical social theory can conceptualize capitalist domination."[10] The totalizing of reification begun in *Dialectic of Enlightenment* leaves "an aesthetic cooperation with nature" as the only domain where "the domination-free interpretation of inner nature is possible."[11]

---

the link between human emancipation and a resurrection of nature. See "Ein philosophierender Intellektueller," *Philosophisch-politische Profile* (Frankfurt am Main: Suhrkamp, 1971), pp. 176–84.

[5] Albrecht Wellmer, *Critical Theory of Society*, trans. John Cumming (New York: Herder and Herder, 1971). Translation of *Kritische Gesellschaftstheorie und Positivismus* (Frankfurt am Main: Suhrkamp, 1969).

[6] Albrecht Wellmer, "Communications and Emancipation: Reflections on the Linguistic Turn in Critical Theory," in *On Critical Theory*, ed. John O'Neill (New York: Seabury Press, 1976), pp. 231–63; quotations from pp. 244, 245.

[7] Axel Honneth, "Communication and Reconciliation: Habermas' Critique of Adorno," *Telos*, no. 39 (Spring 1979): 45–61; quotations from pp. 46–7. The longer German version of this essay was written in 1976, and has been retranslated as "From Adorno to Habermas: On the Transformation of Critical Theory," in Axel Honneth, *The Fragmented World of the Social: Essays in Social and Political Philosophy*, ed. Charles W. Wright (Albany: State University of New York Press, 1995), pp. 92–120.

[8] Friedemann Grenz, *Adornos Philosophie in Grundbegriffen. Auflösung einiger Deutungsprobleme* (Frankfurt am Main: Suhrkamp, 1974).

[9] Thomas Baumeister and Jens Kulenkampff, "Geschichtsphilosophie und philosophische Ästhetik. Zu Adornos 'Ästhetischer Theorie,'" *Neue Hefte für Philosophie*, no. 5 (1973): 74–104.

[10] Honneth, "Communication and Reconciliation," p. 47.   [11] Ibid., p. 50.

Although Honneth subsequently modifies this portrait of Adorno as gloomy aesthete, the main lines to his critique remain the same: *Dialectic of Enlightenment* reflects Adorno's being "imprisoned to a totalized model of the domination of nature" that renders him unable "to comprehend the 'social' in societies."[12] In fact, Honneth sees Adorno, not Horkheimer, as masterminding this fatal collapse of critical social theory.[13] It is primarily Adorno who turns "the interdisciplinary analysis of society" into "the subordinate auxiliary to an aporetic critical theory vacillating between a negativistic philosophy and philosophical aesthetics."[14] Even Seyla Benhabib, who sees more potential than Honneth does in Adorno's concept of "mimetic reconciliation," regards *Dialectic of Enlightenment* as the book that sent Critical Theory into an aesthetic tailspin. She claims that Adorno and Horkheimer seek "the alternative to identity logic and to the domination of internal and external nature in the aesthetic realm." This "turn to the aesthetic" is "astonishing," she says, because "the aesthetic realm offers no real negation of identity logic."[15]

The most forceful statements of such objections, however, have come from Habermas, whom Wellmer, Honneth, and Benhabib regard as having freed Critical Theory from cyclopean bondage.[16] Habermas elaborates his criticisms of *Dialectic of Enlightenment* in two contexts. First, in the final chapter of volume 1 to *The Theory of Communicative Action*, Habermas tries to show why his own account of societal rationalization is a worthy and necessary successor to Western Marxist accounts of reification. There he argues that *Dialectic of Enlightenment* drives Critical Theory into a dead end from which only a "*change of paradigm* within social theory" can deliver it (*TCA* 1:366/489). A paradigm shift is required in order to rescue Critical Theory from Horkheimer and Adorno's fatal expansion of Weber's

---

[12] Axel Honneth, *The Critique of Power: Reflective Stages in a Critical Social Theory*, trans. Kenneth Baynes (Cambridge, Mass.: MIT Press, 1991), p. xii.
[13] Ibid., pp. 35–7.   [14] Ibid., p. 58.
[15] Seyla Benhabib, *Critique, Norm, and Utopia: A Study of the Foundations of Critical Theory* (New York: Columbia University Press, 1986), p. 220.
[16] This is not to deny that these authors have significant criticisms of Habermas, but simply to indicate that all three see his theory of communicative action as a necessary and fruitful advance that extracts Critical Theory from the abyssal aporias in *Dialectic of Enlightenment*.

rationalization thesis, via Lukács's theory of reification, into a totalizing "critique of instrumental reason": a shift from the philosophy of consciousness to a philosophy of communication. According to Habermas, Lukács's theory problematically relied on Hegel's objective idealism, and it was disconfirmed by the failure of proletarian revolutions. This led Horkheimer and Adorno "to sink the foundations of reification critique still deeper and to expand instrumental reason into a category of the world-historical process of civilization as a whole, that is, to project the process of reification back behind the capitalist beginnings of the modern age into the very beginnings of hominization" (*TCA* 1:366/489).

Their theoretical totalizing of reification has three consequences, says Habermas, all of which he aims to avoid: it undercuts the normative foundations for Critical Theory (*TCA* 1:374–7/500–5); it forces Critical Theory to appeal to a concept of truth that it cannot thematize (*TCA* 1:382–3/511–13); and it leads Adorno to renounce social theory altogether in favor of mimetic "gesticulation" (*TCA* 1:385/516). Indeed, Adorno's sole normative foundation comes "shockingly close" to the archenemy Martin Heidegger: "As opposed as the intentions behind their respective philosophies of history are, Adorno is in the end very similar to Heidegger as regards his position on the theoretical claims of objectivating thought and of reflection: The mindfulness [*Eingedenken*] of nature comes shockingly close to the recollection [*Andenken*] of being" (*TCA* 1:385/516).

A similar response to this last theme occurs in *The Philosophical Discourse of Modernity*, the second context where Habermas elaborates his criticisms of Horkheimer and Adorno's alienated masterpiece. Quoting from *Dialectic of Enlightenment*, Habermas says Adorno's *Negative Dialectics* points insistently and relentlessly toward "the prospect of that magically invoked 'mindfulness of nature in the subject in whose fulfillment the unacknowledged truth of all culture lies hidden'" (*PDM* 119–20/145). Now what was "shockingly close" to Heidegger's "recollection of being" is "magically invoked" as well.

This telling phrase occurs in an essay where Habermas compares *Dialectic of Enlightenment* with Nietzsche's critique of the Enlightenment. The comparison is supposed to "forestall" any confusion between Horkheimer and Adorno's "hope of the hopeless" and postmodern "moods and attitudes" emanating "under the sign of a

Nietzsche revitalized by poststructuralism" (*PDM* 106/130). By the end of the essay, however, the similarities to Nietzsche, not the differences, stand out. Habermas identifies three similarities. Like Nietzsche, Horkheimer and Adorno do not do justice to "the rational content of cultural modernity" made available in the differentiation of value spheres and expert cultures (*PDM* 112–14/136–8, 121/146); they undertake a critique of ideology critique that detaches critique from "its own foundations" and regards these foundations "as shattered" (*PDM* 116–19/140–4); and they pursue this totalizing critique from the perspective of "aesthetic modernity," as distinct from scientific or moral modernity (*PDM* 122–5/147–52, 128–9/155–6). Consequently, and again like Nietzsche, Horkheimer and Adorno fall into a "performative contradiction": they use reason to critique the ideological corruption of all reason. Unlike Nietzsche, however, they do not seek refuge in a theory of power, refusing instead to develop any theory to overcome "the performative contradiction inherent in an ideology critique that outstrips itself" (*PDM* 127/154). Adorno in particular strikes Habermas as admirably consistent, for Adorno remains inside this performative contradiction and shows why it cannot be escaped (*PDM* 119–20/145). Nevertheless, Habermas rejects the "purist intent" behind the Frankfurters' hypercritique. It prevents their developing "a social-scientific revision of theory," lands them in "an uninhibited scepticism regarding reason," and forecloses on the possibility that an always already impure reason can nevertheless "break the spell of mythic thinking" without losing touch with "the semantic potentials also preserved in myth" (*PDM* 129–30/156–7).

4.2 REMEMBRANCE OF NATURE

Given such a wide array of serious charges from Adorno's own successor, one wonders why *Dialectic of Enlightenment* remains an important text for Habermas. But one also wonders which text he has read.[17] Other readers might barely recognize his portrait of it as a

---

[17] Habermas's conflicted relation to this text is captured by the remark with which he concludes his interview with Josef Früchtl in 1985: "Even in *Dialectic of Enlightenment* the impulse of the Enlightenment is not betrayed." Jürgen Habermas, *Autonomy and*

critique of instrumental reason that, cut lose from its normative foundations and making wild aesthetic gestures, drifts into the maelstrom of performative self-contradiction. The alternative reading in this chapter singles out two issues on which Habermas and Adorno disagree. I provide my own interpretation of *Dialectic of Enlightenment* with respect to those issues and propose a critical retrieval that does not replicate central problems in Habermas's theory of communicative action. The first issue concerns the scope and imbrication of domination. The second pertains to connections between economic exploitation and societal differentiation. In response, I introduce the idea of "differential transformation" as a post-Habermasian renewal of Adorno's social vision for an age of globalization. My approach will emerge from a selective commentary on "The Concept of Enlightenment,"[18] Horkheimer and Adorno's first chapter on which Habermas is strangely silent.[19]

Unyielding Theory

The difference between Habermas's critical interpretation and my own shows up around the sentence from which he quotes concerning the "mindfulness of nature." Jephcott translates it as follows: "Through this remembrance of nature [*Eingedenken der Natur*] within

---

Solidarity: Interviews with Jürgen Habermas, ed. Peter Dews, rev. ed. (London: Verso, 1992), p. 222.

[18] In one of many noteworthy revisions from the hectographic edition of 1944 to the published edition of 1947, the title of this chapter changed from "Dialectic of Enlightenment" to "The Concept of Enlightenment." For a detailed discussion of such revisions and their significance, see the "Editor's Afterword," *DE* 217-47/423-52.

[19] In fact, the section in *TCA* where Habermas takes *Dialectic of Enlightenment* as his "point of reference" for discussing Horkheimer and Adorno's "reception of Weber" (*TCA* 1:450n2/489n2) relies much more heavily on Horkheimer's *Eclipse of Reason* than on *Dialectic of Enlightenment*. Perhaps that helps explain why Habermas titles this section "The Critique of Instrumental Reason," even though the term "instrumental reason" does not occur in *Dialectic of Enlightenment* and does not capture the target of Adorno's negative dialectic. Significantly, this section in *TCA* cites only two passages from "The Concept of Enlightenment"; see *TCA* 1:379/508 (453n45) and 384/514 (454n57). The second, about "mindfulness of nature in the subject," is cited again in *PDM* 119-20/145. In addition *PDM* cites three other passages from "The Concept of Enlightenment." Neither *TCA* nor *PDM* offers much analysis of what these passages say.

the subject, a remembrance which contains the unrecognized truth of all culture, enlightenment is opposed in principle to power [*Herrschaft*], and even in the time of Vanini the call to hold back enlightenment was uttered less from fear of exact science than from hatred of licentious thought, which had escaped the spell of nature by confessing itself to be nature's own dread of itself" (*DE* 32/64). Habermas and his followers interpret this "mindfulness" or "remembrance" of nature as a "magically invoked" aesthetic relation to nature, a relation that can only be juxtaposed to a totally instrumental rationality. I find little textual basis for their interpretation. The "remembrance" in question is not an aesthetic gesture. It is a process of critical self-reflection no less conceptual and "rational" than any critical social theory that Habermas would endorse. Further, what enlightenment remembers or recollects is not nature in its amorphous nonidentity but rather nature as that which takes distance from itself in fear of its own power, and thereby blindly embraces power.

My interpretation is borne out by the surrounding sentences. Horkheimer and Adorno claim that each "advance of civilization" renews both domination and "the prospect of its alleviation." This prospect depends, they say, "on the concept" – not on aesthetic conduct. Why does it depend on the concept? Because, as good Hegelians, Horkheimer and Adorno think that the concept is inherently critical and self-reflective.[20] As "the self-reflection of thought," the concept can measure the distance that economically driven sciences foster between human beings and nature, a distance that "perpetuates injustice" (*DE* 32/63–4). Accordingly, to "remember nature" in this context is nothing other than conceptually "to recognize power [*Herrschaft*] even within thought as unreconciled nature" (*DE* 32/64). Such conceptual recognition is precisely what Horkheimer and Adorno attempt in their book. Nor do they think conceptual recognition stands outside the scientific enterprise.

---

[20] J. M. Bernstein, *Adorno: Disenchantment and Ethics* (Cambridge: Cambridge University Press, 2001) gives an instructive elaboration of Adorno's emphasis on "the concept," with a view to developing a modernist ethics. On the roots of this emphasis in Kant and Hegel, see Brian O'Connor, *Adorno's Negative Dialectic: Philosophy and the Possibility of Critical Rationality* (Cambridge, Mass.: MIT Press, 2004).

Rather, it is a hidden and suppressed dimension of science itself, without which science loses its emancipatory potential. Although in a "social context which induces blindness" science becomes an instrument whereby people embrace the status quo, science could just as readily let enlightenment fulfill itself by daring "to abolish [*aufzuheben*] the false absolute, the principle of blind power [*Herrschaft*]. The spirit of such unyielding theory would be able to turn back from its goal even the spirit of pitiless progress" (*DE* 33/65).

To understand the confidence Horkheimer and Adorno place in "the concept" and in "unyielding theory," one needs to recall three earlier passages where embers of redemption flicker. Each section in "The Concept of Enlightenment" has such a passage. Together they set a stage for the concluding pages from which I have just quoted. In the first section (*DE* 1–12/25–39), which traces the origins of enlightenment in myth, the diremptions that make enlightenment possible are considered inherently dialectical (*DE* 10–11/37–8). They are inherently dialectical because, in its preanimist origins, language already "expresses the contradiction" that something "is at the same time itself and something other than itself, identical and not identical": a tree addressed as a location of *mana* is both a tree and not a tree. Bound up with language, the concept too originates as "a product of dialectical thinking, in which each thing is what it is only by becoming what it is not" (*DE* 11/37–8). Hence, enlightenment goes wrong not by originating in myth but by suppressing its own dialectical character, in fear of nature's power. So long as enlightenment continues along the path of "mythical fear radicalized," it duplicates rather than recognizes the power of nature: "But this dialectic remains powerless as long as it emerges from the cry of terror, which is the doubling, the mere tautology of terror itself" (*DE* 11/38). From the circle of power and fear arises the fundamentally unjust principle of equivalence (*Prinzip der Gleichheit*, *DE* 12/39) that eventually governs monopoly capitalism as the principle of exchange (*Tauschprinzip*). Already here, then, Horkheimer and Adorno point to a countervailing tendency within enlightenment – dialectical thinking – and a normative basis for their critique of Western society – justice that would "originate in freedom" (*DE* 12/39) rather than in fear.

The second section (*DE* 12–22/39–52) maps an enlightenment path from symbolic myths through cultural separations to mythicized scientific symbols (such as mathematical formulas and fetishized facts). In this context the authors introduce "determinate negation" as an enlightening alternative to both myths and symbols (*DE* 17–18/46–7). Whereas mythic symbols fuse not only sign and image but also reference and referent (*DE* 12/39–40), the mythicized symbols of positivist science and capitalist culture both isolate the sign and eliminate reference. In this way they render "the powerless" mute and "the existing order" unassailable: "Such neutrality is more metaphysical than metaphysics" (*DE* 17/45). The alternative suggested by Horkheimer and Adorno is neither to deny nor to hypostatize the enlightening separation between sign and image (and the concomitant separation between science and art, *DE* 12–13/40), and neither to deny nor to endorse the admixture of oppression and sociality "precipitated in intellectual forms" (*DE* 16/44). They point instead to the dialectical path of "determinate negation." According to their allegorical interpretation, this path opens in the *Bilderverbot* of the Jewish religion (*DE* 17/46). While rendering concepts expressive, determinate negation "discloses each image as script," so that language "becomes more than a mere system of signs" (*DE* 18/46–7). This implies a continual effort to elicit the unfinished meaning of the referent, to grasp "existing things as such ... as mediated conceptual moments which are only fulfilled by [unfolding] their social, historical, and human meaning" (*DE* 20/49). Without the authors explicitly saying this, such mediation of image and concept and such eliciting of unfinished meaning are precisely what *Dialectic of Enlightenment* attempts, in its peculiarly fragmentary way. Far from being the self-contradictory performance of gloomy aesthetes, it is a risk-taking foray into dialectical social criticism, without the Hegelian safety net of "totality ... as the absolute" (*DE* 18/47).

The third section (*DE* 22–34/52–66) diagnoses the societal shape of enlightenment-turned-mythical, especially the impact of monopoly capitalism on labor and human consciousness. There Horkheimer and Adorno make an extraordinary claim: domination is inherently self-limiting, they say, due to the tools it requires (*DE* 29–32/60–3). This claim presents a variation on the Hegelian dialectic of master and slave that organizes the entire section and gets refigured in the commentary on Homer's *Odyssey*, book XII

(*DE* 25–9/55–60).²¹ Whereas in Hegel's struggle for recognition the slave achieves freedom through working upon nature under the master's power, Horkheimer and Adorno suggest that, despite the experiential impoverishment of both "master" and "slave" in a capitalist society, the instruments through which power operates resist being controlled by the ruling class: "The instruments of power [*Herrschaft*] – language, weapons, and finally machines – which are intended to hold everyone in their grasp, must in their turn be grasped by everyone" (*DE* 29/60). The thought here seems to be twofold. First, the purposive character of instruments gives rise to questions about ends that go beyond the restricted "goals" toward which profiteering points them.²² Second, as machines, instruments open up a horizon in which neither forced labor nor class division would be necessary.²³ Far from mounting an all-out assault on instrumental reason, then, passages like this suggest that instrumentalization is itself a potential source of liberation.

Placed side by side, these three passages explain why Horkheimer and Adorno put their confidence in "unyielding theory." For if the concept is dialectical in its origins, if determinate negation provides a conceptual alternative to neutralized rationality, and if the instruments of domination themselves resist domination, then the "disaster" (*Unheil*) radiating across the "enlightened earth" is not inevitable. If it is not inevitable, then a dialectical critique of enlightenment makes sense. In fact, the real antipodes to such a critique

---

²¹ It would be worthwhile to read all of "The Concept of Enlightenment" as a refiguring of themes from Hegel's *Phenomenology of Spirit*. See in this connection the highly illuminating essay by J. M. Bernstein, "Negative Dialectic as Fate: Adorno and Hegel," in *The Cambridge Companion to Adorno,* ed. Tom Huhn (Cambridge: Cambridge University Press, 2004), pp. 19–50. Bernstein sees Adorno's philosophy as rearticulating "what it means to be a Hegelian ... after two centuries of brutal history in which the moment to realize philosophy ... was missed" (p. 20), and he interprets *Dialectic of Enlightenment* as "a generalization and radicalization" of the dialectic between pure insight and religious faith in the chapter on "The Enlightenment" in Hegel's *Phenomenology of Spirit* (p. 22).
²² In the authors' words: "The thing-like quality of the means, which make the means universally available ..., itself implies a criticism of the domination from which thought has arisen as its means" (*DE* 29/60).
²³ As Horkheimer and Adorno put it: "In the form of machines, however, alienated reason is moving toward a society which reconciles thought, in its solidification as an apparatus both material and intellectual, with a liberated living element, and relates it to society itself as its true subject" (*DE* 29/60–1).

are not positivism and the culture industry, but political and economic rulers who desperately attempt, at this late stage, to pose "as engineers of world history" (*DE* 30/61). It would not be farfetched to regard the rulers' desperation as a hopeful sign: not even those who benefit most from class-based domination believe in its necessity.

This makes it easier to grasp why necessity does not have the last word in thought either. Given its dialectical character, thought, as an "idea-tool," can rightly insist on separations, say, between subject and object, yet, when applied to itself, can also recognize separations as an "index" both of thought's own untruth and of truth. Proscribing the fusions of superstition, enlightenment has always both promoted and opposed domination. And from this the "remembrance of nature" flows. Let me quote at some length: "Enlightenment is more than enlightenment, it is nature made audible in its estrangement. In mind's self-recognition as nature divided from itself, nature, as in prehistory, is calling to itself, ... no longer directly by its supposed name, ... but as something blind and mutilated. In the mastery of nature, without which mind does not exist, enslavement to nature persists. By modestly confessing itself to be power [*Herrschaft*] and thus being taken back into nature, mind rids itself of the very claim to mastery [*der herrschaftliche Anspruch*] which had enslaved it to nature" (*DE* 31/63). For this, only unyielding theory will suffice, theory that neither mistakes existing institutions and practices "for guarantors of the coming freedom" (*DE* 31/63) nor succumbs to nature's thrall by accepting the necessity of blind domination.

Domination and Exploitation

In Chapter 2 I expressed reservations about Adorno's objectification of transformative hope. There I suggested two reasons why Adorno bases his hope for societal transformation upon nonidentical objects: because he totalizes transformation, and because he fails to distinguish sufficiently between societal evil and the violation of societal principles. Habermas has the same tendencies in view when he takes issue with what he sees as a theoretical totalizing of reification in *Dialectic of Enlightenment* and in Adorno's subsequent writings. Yet I do not think that Adorno actually totalizes reification in the way

Habermas describes, nor does Adorno put the emphasis on reification that Habermas alleges. Adorno's *Negative Dialectics* very clearly calls reification "an epiphenomenon," compared with "the possibility of total catastrophe" (*ND* 190/191).[24] The theoretical moves that lead to an apparent impasse in Adorno's critical social theory cannot be explained along Habermasian lines. They have to do, rather, with the two issues I wish to examine: the scope and imbrication of domination, and connections between economic exploitation and societal differentiation. To set the stage, permit me two remarks of a systematic nature. The first has to do with the much disputed relation between subject and object. The other has to do with the status of normative judgments about economic systems.

Adorno's insistence in *Negative Dialectics* on "the priority of the object" and Habermas's relative neglect of this Adornian theme are equally noteworthy. The point to Adorno's insistence is both to recall normative limits to the subject's ability to "constitute" the object and to remind us that the subject is itself an object at its core. Missing in Adorno's account, however, and even more strikingly absent in Habermas's theory of communicative action, is any indication that the object can also be a subject. Both Adorno and Habermas omit this because they have epistemological notions of "subject" and "object." That is to say, they define the object primarily as what human beings can know and, by extension, what they can (try to) make, control, or influence.[25] But this leaves out of account an entire range of relations within which human beings and their "objects" are mutual subjects.

---

[24] For a thorough commentary on this remark that shows why Habermas's interpretation of Adorno is mistaken and his "linguistic turn" is problematic, see the master's thesis by Matthew J. Klaassen, "The Nature of Critical Theory and Its Fate: Adorno vs. Habermas, Ltd." (Institute for Christian Studies, Toronto, 2005). Steven Vogel also emphasizes Adorno's reservations about Lukács's theory of reification, but he claims that Adorno's alternative relies on a highly problematic appeal to immediacy. It seems to me that Vogel's claim echoes Habermas's failure to understand the Hegelian dimensions to Adorno's "priority of the object." See Steven Vogel, *Against Nature: The Concept of Nature in Critical Theory* (Albany: State University of New York Press, 1996), pp. 69–90.

[25] Prior to the publication of *TCA*, Joel Whitebook, "The Problem of Nature in Habermas," *Telos*, no. 40 (Summer 1979): 41–69, had already identified tendencies that make it very difficult for Habermas to regard nonhuman creatures as anything other than epistemological and instrumental objects. Whitebook's essay points up the need to find a more viable way to address the "ecological crisis," but it does not provide a significant alternative to Habermas's approach.

Animals, for example, perceive us just as much as we perceive them, and they have needs and emotions that no mere "object" could have. So too, humans share biospheres with plants and animals. Although dramatically shaped by human activity, for better and for worse, biospheres are co-constituted by nonhuman life. In that sense plants and animals have an "agency," or at least a subjectivity, that exceeds mere "objecthood," and on which human "subjects" depend. The relevance of this remark for the question of domination will become apparent shortly.

Closely connected to the disputed relation between subject and object is the question of normative judgments about economic systems. Is it appropriate to evaluate and criticize an economic system in terms of societal principles such as solidarity and justice and in terms of an economy's contribution to human flourishing? Ambivalence about this question in Critical Theory goes back to Hegel and Marx. It surfaces, for example, in ongoing debates about whether there are "moral" underpinnings to Marx's critique of capitalism. Adorno's resistance to the "exchange society" implies a critique of economic exploitation and of the needless suffering it creates. To that extent *Dialectic of Enlightenment* and *Negative Dialectics* suggest, but do not elaborate, normative judgments about capitalism as an economic system, from the perspective of "damaged life." Habermas seems less concerned about damage that is internal to the capitalist economy. Under the theme "colonization of the lifeworld," he shifts the focus of normative critique outward, to the economic system's impact on noneconomic practices and institutions.[26] It seems to me, by contrast, that neither implicit nor external normative judgments suffice. For if the nerve center of societal evil in its modern form lies in a capitalist economy, then a philosophical diagnosis needs to be explicit in its normative critique of the economy as such. What this entails will emerge a little later.

Both the subject-object relation and the question of normative critique are at work in "The Concept of Enlightenment." This can be seen from the prominence given to a pattern of blind domination

---

[26] For a comparison of Habermas and Adorno that is critical of Habermas on this score, see Deborah Cook, *Adorno, Habermas, and the Search for a Rational Society* (New York: Routledge, 2004), pp. 39–70.

when Adorno and Horkheimer explain the "disaster triumphant" that has befallen "the wholly enlightened earth." In their account, blind domination occurs in three tightly interlinked modes: as human domination over nature; as domination over nature within human existence; and, within both of these modes, as the domination of some human beings by others. To provide terminological markers for these three modes of domination, I shall use the terms "control," "repression," and "exploitation," respectively. Critics of Adorno either downplay one of these modes or argue that they are not tightly interlinked in the manner he suggests. My own response is that all three modes do actually characterize modern Western societies and that understanding their interlinkage is crucial for a transformative social theory. But I want to propose that each mode of domination deserves its own form of normative critique. *Dialectic of Enlightenment* hovers near the trap of a totalizing critique precisely because it does not differentiate sufficiently in its critique of domination. Although Habermas has called attention to insufficient differentiation, he does so at the cost of ignoring continuities among control, repression, and exploitation.

According to *Dialectic of Enlightenment*, violence is systemic in modern Western societies. This systemic violence has emerged in a specific configuration, namely, in the imbrication of control (*Naturbeherrschung*) with repression and exploitation. Further, Western exploitation involves a class-based division of labor that traces back to wars of territorial conquest and the establishment of a social order based on fixed property (*DE* 9/36). The differentiation of cultural spheres, and particular advances within science, art, and morality, are neither separate from nor reducible to such societal tendencies. In order for all of these developments to deliver what they promise – for so-called progress not to be cursed with "irresistible regression" (*DE* 28/59) – systemic violence needs to be recognized and resisted. That, in my own language, is the truth to Adorno's "remembrance of nature," and a blind spot in Habermas's critique.

Yet systemic violence becomes difficult to recognize and resist if control, repression, and exploitation become fused in the critical concept of "domination." Let me begin with the notion of control. Adorno does not consider all control of "nature" to be illegitimate. In fact, he regards some control to be necessary if human freedom is to be possible. So one wonders how the distinction should be drawn between legitimate and liberating control, on the one hand, and illegitimate and destructive control, on the other. In other words, what is askew in the enlightenment vision of "liberating human beings from fear and installing them as masters"? Adorno suggests that enlightened mastery gets distorted in being driven by a fear of nature's power. But what would be the alternative to fear? Presumably it would be a form of recognition, as Adorno's *Eingedenken der Natur* suggests. Yet it cannot be a straightforward recognition of "nature" as "other" than human, nor can it be merely a recognition of nature's power as the object of fear. The first would not support discriminations between legitimate and illegitimate control, and the second would provide little basis for distinguishing liberation from destruction.

The recognition required pertains to the mutual intersubjectivity of human beings with other creatures in the dimensions of life they share. It pertains, for example, to the mutual interdependency of all organisms in the biospheres they inhabit. Human control of other organisms becomes illegitimate when it no longer promotes their interconnected flourishing. It becomes destructive when it promotes human flourishing at the expense of all other organisms. Although it is not necessarily illegitimate and destructive for human beings to treat other organisms as objects in dimensions they do not share, to treat them as no more than objects and to ignore the dimensions of subjectivity we share with them are forms of violence. Unfortunately, as Horkheimer and Adorno suggest, such violence has permeated Western civilization from its earliest stages. To base the pursuit of human freedom on misrecognition of other creatures is to violate the very meaning of freedom. Freedom is not a freedom to dominate but a freedom to flourish.

As a second mode of domination, repression is closely linked to the first. But the two differ in significant respects. Adorno's excursus on the *Odyssey* describes repression as the introversion of sacrifice. The emergent self attains its masterful identity by sacrificing its own happiness:

> The identical, enduring self which springs from the conquest of sacrifice is itself the product of a hard, petrified sacrificial ritual in which the human being, by opposing its consciousness to its natural context, celebrates itself. ... In class society [*Klassengeschichte*], the self's hostility to sacrifice

included a sacrifice of the self, since it was paid for by a denial of nature in the human being for the sake of mastery [*Herrschaft*] over extrahuman nature and over other human beings. ... The human being's mastery [*Herrschaft*] of itself ... practically always involves the annihilation of the subject in whose service that mastery is maintained. ... The history of civilization is the history of the introversion of sacrifice – in other words, the history of renunciation. (*DE* 42–3/77–9)

Yet Adorno does not think that all self-mastery is illegitimate, or that all delayed gratification is destructive. Sensuous nirvana is not the same as human freedom. In what does the difference consist?

For Adorno the crucial difference seems to lie between repression and sublimation. In repression, our urges and desires are ignored or denied. In sublimation, they are accepted and redirected. But the accepting and redirecting of urges and desires raises a normative question that Adorno avoids: what makes our urges and desires acceptable and worthy of redirection? To answer requires an account of basic needs, not merely my needs or yours, but needs whose satisfaction, with suitable cultural inflections, would characterize any human flourishing. Among these would be needs for adequate food and drink, for shelter and clothing, for affection and companionship, and the like. On my account, those urges and desires are acceptable and worthy of redirection which lead to the satisfaction of such basic needs. Moreover, because their satisfaction rarely occurs outside the context of other human beings, a culture that represses our urges and desires is just as problematic as a societal formation that grants satisfaction on the part of some human beings by denying it to others.[27]

---

[27] To relate my account of sublimation to Adorno's more cautious approach would require a lengthier discussion, especially of two manuscripts that Adorno wrote in 1942 when he and Horkheimer were working on *Dialectic of Enlightenment*. Titled "Reflexionen zur Klassentheorie" and "Thesen über Bedürfnisse" [Theses about Needs], they were published posthumously in the collection *Soziologische Schriften I*, *Gesammelte Schriften* 8 (Frankfurt am Main: Suhrkamp, 1972), pp. 373–91 and 392–6, respectively. The first has been translated as "Reflections on Class Theory," in *Can One Live after Auschwitz? A Philosophical Reader*, ed. Rolf Tiedemann, trans. Rodney Livingstone et al. (Stanford: Stanford University Press, 2003), pp. 93–110. For a different and more psychoanalytic account of sublimation, see Joel Whitebook, *Perversion and Utopia: A Study in Psychoanalysis and Critical Theory* (Cambridge, Mass.: MIT Press, 1995), especially pp. 217–62. I share Whitebook's dissatisfaction both with Adorno's failure to spell out a theory of sublimation

Such a society would be exploitative. According to *Dialectic of Enlightenment*, the various societal formations that have characterized Western civilization, from Homeric times to the twentieth century, have been exploitative. Further, the advances brought about by the process of enlightenment have occurred not *despite* exploitation but *by way of* exploitation.[28] By "exploitation" I mean a one-sided social distribution of power in which the apparent flourishing of one group persistently occurs at the expense of another. Although such a distribution need not require a class-based division of labor, I accept Adorno's Marxist claim that it has been class based in Western societies. I also share Adorno's intuition that exploitation is always illegitimate and destructive, directly destructive for the exploited and indirectly destructive for the exploiters. I would assert more explicitly than Adorno does, however, that the reason why exploitation is illegitimate is that it violates fundamental principles of solidarity and justice without which societal freedom cannot be attained.

Accordingly, I distinguish three forms of violence in Western societies and posit a distinct normative basis for criticizing each. The control of nature becomes violent when it does not promote the interconnected flourishing of all creatures but promotes human flourishing at the expense of all other creatures. The formation of the self becomes violent when it represses urges and desires that would lead to the satisfaction of basic needs. And the social distribution of power becomes exploitative, and therefore illegitimate and destructive, when it persistently promotes the apparent flourishing of one group at the expense of another.

---

and with Habermas's failure to take seriously the need for such a theory. Whitebook argues convincingly that the demise of psychoanalytic reflection has direct links to the "de-utopianization" and "decorporealization" of Habermasian Critical Theory.

[28] The term "exploitation," used frequently in the 1944 hectograph, is replaced by less loaded terms such as "enslavement" in the 1947 published version. For the theoretical debates about "state capitalism" informing this change, see Willem van Reijen and Jan Bransen, "The Disappearance of Class History in 'Dialectic of Enlightenment': A Commentary on the Textual Variants (1947 and 1944)," in *DE* 248–52/453–7. I have revived the term "exploitation" because it more precisely indicates the societal mode of domination that Horkheimer and Adorno have in mind.

4.3 BEYOND GLOBALIZATION

How should we understand the interlinkage among these three modes of domination? If Adorno does not totalize the domination of nature (contra Honneth) and does not project reification back into the dawn of humanity (contra Habermas), and if such a collapsing of distinctions cannot in any case support a transformative social theory, how should we thematize the connections among control, repression, and exploitation? My response is this. Although exploitation, as it has taken shape in modern Western societies, requires a high degree of repression and destructive control, repression and destructive control are not peculiar to Western societal formations. Nor would they automatically disappear if Western-style exploitation ended. Yet, because capitalism as an economic system depends so heavily on patterns of repression and destructive control, it would be very difficult to disrupt these patterns today if economically anchored exploitation were not dismantled.

Normative Critique

To envision a dismantling of exploitation, a transformative social theory needs to include a normative critique of capitalism as an economic system. This critique would have two themes. First, it would need to come to grips with the totalizing character of the "logic" of capitalism. Despite the pitfalls introduced by Adorno's critique of "identity thinking" and the "exchange society," he has recovered Marx's insight into the inexorably expansive character of a capitalist economy. Such an economy is not simply one system within a larger and differentiated societal whole, as Habermas often seems to suggest. It inherently tends to dominate the whole, an expansionary drive that *Minima Moralia* succinctly captures in Adorno's famous parody of Hegel: "The whole is the false" (*MM* §29, p. 50/55).[29] To stop growing would eventually spell the end of capitalism. The commodification of culture, the militarization of space, and the perpetual destruction of biospheres are simply different manifestations of this

---

[29] Hegel's dictum was "The True is the whole." See G. W. F. Hegel, *Phenomenology of Spirit*, trans. A. V. Miller (Oxford: Oxford University Press, 1977), p. 11.

systemic imperative at work. It is a real question whether capitalism as such leaves room in the long term for "sustainable development."

The imperative to expand presupposes a second troubling feature of capitalism, namely, its channeling intrinsically collective and public goods into private and privileged pockets. Despite problematic aspects to Marx's labor theory of value, his account of "surplus value" identifies this feature with remarkable clarity and insight. In nontechnical language, the secret of capitalism, which mainstream economic theories occlude, is that it must continually generate excess returns for those who occupy positions of economic power, whether they be individual investors, transnational corporations, or the most prosperous countries in the world economy. Attempts to rectify resulting imbalances in the distribution of wealth – charity, progressive taxation, debt relief, foreign aid, and the like – do not challenge the continuation of this inherently exploitative system.

It is one thing to identify the totalizing and exploitative character of capitalism, however, and quite another to say how the economic system should be transformed. Recognizing this difference, Adorno has little to say about the sorts of changes needed. Yet he insists on the historical possibility of such a transformation, and he consistently indicates that a postcapitalist economy will need to be neither totalizing nor exploitative. It will also need to be an economy that does not heavily depend upon repression and destructive control. In my own terms, what is needed is a differential transformation of Western society.[30]

Differential Transformation

By "differential transformation" I mean a process of significant change in contemporary society as a whole that occurs at differing levels, across various structural interfaces, and with respect to distinct

---

[30] From the description of "differential transformation" that follows, it should be clear that, like Max Pensky, I recognize three challenges facing an attempt to "globalize critical theory" in order to theorize globalization: to develop a highly reflexive social theory, to pursue a new form of interdisciplinarity, and to explicate the socially embedded grounds of normative critique. See his "Globalizing Theory, Theorizing Globalization: Introduction," in *Globalizing Critical Theory*, ed. Max Pensky (Lanham, Md.: Rowman & Littlefield, 2005), pp. 1–15.

societal principles. The primary levels in question are social institutions, cultural practices, and interpersonal relations. The primary structural interfaces lie among economy, polity, and civil society. The most relevant societal principles in this context are those of resourcefulness, justice, and solidarity. Given the tight links among exploitation, repression, and destructive control, no single societal site can suffice as an arena in which to promote creaturely flourishing. Yet changes in many diverse sites also will not suffice if they do not move in mutually reinforcing directions. Hence, the transformation of society as a whole needs to be an internally differentiated and complementary process. The differentiation of levels and principles in Western society provides a historical basis for such a process.

Seen from this perspective, the debate between advocates and opponents of "globalization" misses some of the central issues at stake. Few would deny that the world is caught up in an "expanding scale, growing magnitude, speeding up and deepening impact of transcontinental flows and patterns of social interaction," such that "distant communities" are more directly linked and "the reach of power relations [expands] across the world's regions and continents."[31] But this does not tell us which direction these changes should take and on what historical basis they are possible. The idea of differential transformation provides a historically informed basis for attempting a normative critique of globalization, in three respects.

First, globalization needs to be evaluated with regard to different levels of social interaction. Two considerations come into play. One is that the tasks of differentiated institutions and practices need to be strengthened rather than undermined, and the intrinsic worth of interpersonal relations cannot be forgotten. Accordingly, patterns of globalization that turn cultures into economic war zones need to be identified and resisted by organizations and agencies within them, in the name of upholding societal differentiation and interpersonal connections. The "no logo" slogan of antiglobalization activists points in this direction, although their anarchist formulations often suggest a communalist and dedifferentiating agenda. The second consideration is that, to maintain and strengthen societal differentiation, economic

---

[31] David Held and Anthony McGrew, *Globalization/Anti-Globalization* (Cambridge: Polity Press, 2002), p. 1.

alternatives to the capitalist juggernaut must be fostered. In the absence of nonprofit, cooperative, and community-based modes for securing resources and providing goods and services, noncommercial organizations and agencies will not thrive. So another basis for evaluating patterns of globalization is the extent to which these permit and promote economic alternatives.

Globalization also needs to be evaluated in terms of the structural interfaces among economy, polity, and civil society. Arguably an achievement of modernization has been to create societal subsystems that follow their own imperatives. The result is a society in which, for example, law and politics are not supposed to serve merely private economic interests, and the organizations and agencies of civil society are not supposed to serve merely the interests of state. Admittedly, the integrity of these subsystems is constantly threatened. The economically and politically powerful regularly subvert it in their pursuit of greater wealth and power. Yet the subsystems remain mostly intact, and their boundaries are relatively clear. This provides a basis for evaluating patterns of globalization. As Habermas's thesis about the "colonization of the lifeworld" suggests, patterns that subsume one subsystem under another are inherently destructive and unstable. The two obvious examples are economic imperialism and political authoritarianism, which recently have joined forces in a volatile and violent American Empire. Clearly, if globalization simply means the spread of this empire, all bets are off. Yet even a critique of globalization as a "new imperialism" needs to go beyond a sort of reverse nationalism.[32] It needs to insist on the integrity of distinct subsystems, and it needs to detect those spots where structural interfaces have been weakened or overridden. For a subsystem cannot maintain its integrity if it does not open properly toward the imperatives of the other subsystems.

Finally, globalization needs to be evaluated with respect to societal principles that are pervasive, distinct, and mutually complementary. I have mentioned three such principles: resourcefulness, justice, and solidarity. Being pervasive, these principles pertain to all of the levels and subsystems of a differentiated society. Yet each of them holds in a special way for a distinct range of levels and subsystems.

---

[32] See David Harvey, *The New Imperialism* (Oxford: Oxford University Press, 2003).

Resourcefulness, for example, has decisive relevance for economic institutions and the economy as a whole; justice, for political institutions and the state; and solidarity, for cultural practices and civil society. The theoretical challenge, and a practical one as well, is to envision normative integration across these differentiated zones without either allowing one to dominate the others or exempting any zone from the requirements of all three principles. Specifically, given the expansionary drive of capitalism, economic institutions and patterns need to demonstrate resourcefulness without promoting injustice and alienation.

Two insights are crucial in this regard. The first is that, for the most part, resourcefulness is not the operative imperative of contemporary capitalist economies. Capitalism distorts the principle of carefully stewarding human and nonhuman potentials for the sake of interconnected flourishing. Capitalism twists this principle in the direction of efficiency, productivity, and maximal consumption for their own sakes. Consequently, considerations of justice and solidarity become economic afterthoughts. They turn into belated attempts to alleviate the damage necessarily done by a system that does not prize resourcefulness in the first place.

The second insight, closely connected to the first, is that a society must follow all three principles at the same time in order to follow any one of them. Dutch economist and social philosopher Bob Goudzwaard calls this "the simultaneous realization of norms."[33] According to Goudzwaard, the industrial revolution and the development of capitalism problematically gave "well-nigh *absolute* priority" to "technical and economic progress" of a certain sort, while turning societal principles such as justice and solidarity into mere means to achieve such "progress." Let me quote a central passage from his critique: "Capitalism is subject to critique insofar as, for the sake of progress, it is founded on independent and autonomous forces of economic growth and technology, that is, forces which are considered isolated, sufficient, and good in themselves. These economic

---

[33] Bob Goudzwaard, *Capitalism and Progress: A Diagnosis of Western Society*, trans. and ed. Josina Van Nuis Zylstra (Grand Rapids, Mich.: Eerdmans, 1979), pp. 65–8, 204–23. Goudzwaard, in turn, derives the idea of a simultaneous realization of norms from the work of economist T. P. van der Kooy (see p. 65n30) and the philosopher and legal theorist Herman Dooyeweerd.

and technological forces are indeed related to norms of ethics and social justice, but in such a manner that these norms cannot impede the realization of those forces and the promotion of 'progress.' These norms are consciously viewed as dependent upon and secondary to the forces of progress: they are placed in the service of the expansion of technology and the growth of the economy."[34] According to Goudzwaard, the instrumentalizing of societal principles and the failure to pursue a "simultaneous realization of norms" have two effects. First, they distort the meaning of economic resourcefulness. Second, they turn all institutions and interpersonal relations into means to "economic" ends. Given the dominance of capitalism, Western society has become a "tunnel society" in which "everything – people, institutions, norms, behavior – contributes to the smooth advance toward the light at the end of the tunnel. But the end of the tunnel never appears to be within reach; the light shines forever *in the future*. Nevertheless, it keeps everything and everyone in the tunnel on the move. . . . *[F]unctional streamlining* [is] imposed on the social order in each of its aspects. Nothing is of essential value in any social relationship unless it is a means to advance in the tunnel."[35]

In my own terms, Goudzwaard points up a central requirement for the differential transformation of society. Differential transformation will not occur unless distinct and pervasive societal principles such as resourcefulness, justice, and solidarity are in effect across the board and are not relegated to separate zones. The intrinsic meaning of each depends on the simultaneous holding of the others. Here "holding" means both that the principles hold for the members of a society and that the members, amid their social struggles and conflicting interpretations, hold these principles in common.[36] In other words, the principles are mutually complementary, and existing societal patterns need to be evaluated in that regard. Although, like Goudzwaard, I find Western societies woefully lacking in this respect, I also judge that modern differentiation creates conditions that would allow greater complementarity in the future. But this is not the occasion to make a detailed case in support of this judgment.

---

[34] Ibid., p. 66; entire passage italicized in the original.   [35] Ibid., pp. 183–4.
[36] For a fuller description of societal principles as commonly holding and commonly held, see Lambert Zuidervaart, *Artistic Truth: Aesthetics, Discourse, and Imaginative Disclosure* (Cambridge: Cambridge University Press, 2004), pp. 96–100.

We can summarize the argument of this chapter as follows. Whereas Habermas rightly criticizes Adorno for missing the potential of modern differentiation to promote human flourishing, Habermas has an inadequate analysis of the systemic violence that accompanies this differentiation. Conversely, whereas Adorno rightly calls attention to systemic violence in its several and interconnected modes, he tends to totalize it in a way that seems to preclude genuine change in a societal formation. On my own account, the change required will be a differential transformation. It will occur across different social institutions, cultural practices, and interpersonal relations. It will rearticulate the structural interfaces among economy, polity, and civil society. And it will ensure that distinct societal principles such as resourcefulness, justice, and solidarity are neither played off against one another nor instrumentalized into means to a single discrete end. Instead, these societal principles will be mutually complementary, not only in the economy, where they are often violated or ignored, but also in the state, in civil society, and, indeed, in society as a whole.

What Adorno articulates more eloquently than his successors is that "the whole is the false." In the long run, we cannot resist the repression of desire and the destruction of nature unless we dismantle economic exploitation. What he needed to say more vigorously, however, and with greater nuance, is that the whole is not wholly false. This is the valid point to Habermas's otherwise overwrought critique. Like the book over which the ways of Critical Theory seem ever to part, the dialectic of enlightenment is an alienated masterpiece. To disalienate both, to turn their mastery toward differential transformation, would be signs of hope in the face of "disaster triumphant."

5

Autonomy Reconfigured

> This is not a time for political works of art; rather, politics has migrated into the autonomous work of art, and it has penetrated most deeply into works that present themselves as politically dead.
>
> Adorno, *Notes to Literature*[1]

Adorno's critique of the culture industry is both highly relevant and historically dated, both theoretically provocative and politically problematic. Developed in the 1930s and 1940s before the rise of new social movements, Adorno's critique identifies crucial issues that contemporary feminism needs to address. Although his critique turns on an idea of artistic autonomy to which many feminists object, their objections make social-theoretical assumptions that Adorno's critique can challenge.

Continuing my efforts at a critical retrieval, this chapter explores the relevance of Adorno's critique for contemporary feminism. After

---

Excerpts from this chapter were presented at a Critical Theory Roundtable in Hayward, California (October 2001) and in a panel on "Socio-Political Issues in Feminism and Aesthetics" at the American Society for Aesthetics (ASA) meeting in Coral Gables, Florida (November 2002). I wish to thank the participants in these sessions for their comments, especially L. Ryan Musgrave, who organized the ASA panel. I also wish to thank Renée Heberle for her instructive comments on an earlier version of this chapter.

[1] Theodor W. Adorno, "Commitment," in *Notes to Literature*, vol. 2, ed. Rolf Tiedemann, trans. Shierry Weber Nicholsen (New York: Columbia University Press, 1992), pp. 93–4; "Engagement," in *Gesammelte Schriften* 11 (Frankfurt am Main: Suhrkamp, 1974), p. 430.

identifying a tension within feminism concerning the idea of artistic autonomy, I examine the role this idea plays in Adorno's critique of the culture industry. I then conclude by proposing a reconfigured notion of artistic autonomy, one that aims to overcome the limitations of Adorno's critique and to support a revitalized feminist politics.

5.1 FEMINIST CULTURAL POLITICS

The idea of artistic autonomy emerged from the eighteenth-century Enlightenment in Europe. It posits that the arts, and the sorts of experience that the arts afford, are properly independent from other types of human endeavor and need to follow their own rules. In the twentieth century, modernist aesthetics made an emphasis on authentic works of art central to this idea of autonomy. Not without criticism, however. The modernist attempt to anchor artistic autonomy in authentic artworks was challenged by avant-garde movements such as dada and surrealism and by a turn toward socially engaged art at both ends of the political spectrum. What I have described as the "paradoxical modernism" of Adorno's aesthetics took shape in this environment.[2] In debate with Walter Benjamin, Bertolt Brecht, and Georg Lukács in the 1930s, Adorno sided with other modernists against the avant-garde and against advocates of political "commitment" in the arts and scholarship.[3] As he puts it in his polemic from the 1960s against political works of art, "politics has migrated into the autonomous work of art."[4]

Adorno's stance looks problematic from a feminist perspective.[5] He seems to endorse the very notion that feminist cultural politics

---

[2] See "Paradoxical Modernism," in Lambert Zuidervaart, *Adorno's Aesthetic Theory: The Redemption of Illusion* (Cambridge, Mass.: MIT Press, 1991), pp. 150–77.
[3] See "Aesthetic Debates," in Zuidervaart, *Adorno's Aesthetic Theory*, pp. 28–43. Key documents in these debates are collected in *Aesthetics and Politics: Debates between Bloch, Lukács, Brecht, Benjamin, Adorno*, ed. Ronald Taylor, with an afterword by Fredric Jameson (London: NLB, 1977). See also Eugene Lunn, *Marxism and Modernism: An Historical Study of Lukács, Brecht, Benjamin, and Adorno* (Berkeley: University of California Press, 1982).
[4] See "Political Migration," in Zuidervaart, *Adorno's Aesthetic Theory*, 122–49.
[5] Perhaps this is one reason why, with notable exceptions such as Seyla Benhabib, Drucilla Cornell, and, more recently, Judith Butler, feminists in North America have paid little attention to Adorno's work. That is starting to change, however, as is

resists, a Western notion of artistic autonomy that, in shoring up a patriarchal culture, excludes women or marginalizes their voices. As Mary Devereaux has shown, feminist resistance aims especially at an autonomist emphasis on formal qualities in art and on formal criteria for evaluating art.[6] The formalism of, say, Clive Bell and Clement Greenberg seems to sever art from its roots in embodied lives and to gut its role in politics and society. Moreover, formalist approaches to art history and art criticism not only misinterpret politically engaged art but also exclude women from the canon of artistic achievement. An autonomist stress on the formal side of art, which Adorno shares, has masculinist implications that feminists have persuasively criticized.

Yet the critique of artistic autonomy also has potential problems from a feminist perspective. While granting the legitimacy of the feminist critique, Devereaux raises concerns that resonate with Adorno's worries about politically engaged art. She says an otherwise beneficial rejection of artistic autonomy runs two risks, one theoretical and the other practical. Theoretically, feminist criticisms of artistic autonomy threaten to ignore or underestimate "the elements that make art *art*." Among such elements, Devereaux, like Adorno, would include formal matters, as distinct from content and context. Practically, feminist criticisms of artistic autonomy might have the unintended consequence of "exposing art to political interference." Devereaux explains:

> Historically, the separation of the aesthetic and the political has provided an argument both against artistic censorship, narrowly defined, and what John Stuart Mill called "the tyranny of the majority." When threatened with interference, artists and their supporters simply appealed to the idea of the

---

signaled by the appearance of *Feminist Interpretations of Theodor Adorno*, ed. Renée J. Heberle (University Park: Pennsylvania State University Press, 2006).

[6] Mary Devereaux, "The Philosophical and Political Implications of the Feminist Critique of Aesthetic Autonomy," in *Turning the Century: Feminist Criticism in the 1990s*, ed. Glynis Carr (Lewisburg, Pa.: Bucknell University Press, 1992), pp. 164–86. This is a special issue of the *Bucknell Review* 36, no.2 (1992). See also Devereaux's essays "Protected Space: Politics, Censorship, and the Arts," *Journal of Aesthetics and Art Criticism* 51 (Spring 1993): 207–15, and "Autonomy and Its Feminist Critics," *Encyclopedia of Aesthetics*, ed. Michael Kelly, vol. 1 (New York: Oxford University Press, 1998), pp. 178–82. For reasons explained in Chapter 1, I use the term "artistic autonomy" for what Devereaux calls "aesthetic autonomy."

"autonomy" of art, claiming the illegitimacy of any evaluative criteria other than the purely aesthetic. But having abandoned strictly "aesthetic" (i.e., formal) criteria in favor of a wider set of political and social considerations, feminist critics of autonomy need a principled basis for distinguishing legitimate from nonlegitimate grounds of evaluation. If a work's misogyny may be relevant to its assessment as art, then why not its failure to promote the "family values" demanded by Senator Jesse Helms and others on the political right?[7]

In a similar vein, when Adorno criticized politically engaged art in the 1960s, he did so in part because such art lent unintended support to a dangerous moral censoriousness that abetted Germany's postwar repression of guilt and suffering.

Adorno would have recognized and refused the dilemma Devereaux poses for aesthetics – in her formulation, "*either* adopt a theory of autonomy that protects art from the exigencies of political fashion but isolates it from life, *or* opt for a political conception of art that integrates art with life at the price of compromising its independence." To avoid this dilemma, Devereaux proposes to redefine autonomy as "the idea that works of art deserve a protected space, a special normative standing." By "protected space" she means that "works of art ... remain under the control of artists and the institutions of the art world in which they work." Artworks deserve such protection for a political reason, she says: "[T]hey often play an important social and political role: pushing beyond or challenging existing ways of seeing and thinking about the world."[8]

This proposed redefinition reminds one of Adorno's describing autonomous art as "the social antithesis of society" (*AT* 8/19) whose critical capacities depend on its relative independence from the rest of society. But Adorno would have questioned Devereaux's emphasis on a "protected space." For he considers late capitalist society as a whole to be much more integrated and much more oppressive than Devereaux assumes. To think that existing practices and institutions of art could secure a "protected space" from "political interference," even if only in a symbolical fashion, would be politically naive and theoretically myopic. Naive, because the dominant institutions of government work hand in glove with an exploitative economy.

---

[7] Devereaux, "Autonomy and Its Feminist Critics," pp. 181–2. [8] Ibid., p. 182.

Myopic, because the government-maintained capitalist system surrounds and permeates the very spaces that are supposed to be protected.

Despite Adorno's emphasis on artistic autonomy, then, the social theory informing his paradoxical modernism resembles the social theories of radical and socialist feminists who have a strong critique of autonomy.[9] Both radical and socialist feminists launch their critiques of artistic autonomy out of opposition to the societal system as a whole – precisely that system to which autonomous art offers an important social antithesis, according to Adorno. Radical feminists think the oppression of women stems from a patriarchal societal system that has either biological or cultural roots. This patriarchal system thoroughly devalues women and their experience. Liberation would require breaking the grip of patriarchy on women's lives. The path to that lies in either subverting the patriarchal system or escaping it by developing a counterculture. Socialist feminists also think that the oppression of women is systemic. They differ, however, in understanding the system as a historical formation in which economic patterns are decisive. Socialist feminists claim that the liberation of women would require a fundamental transformation of the patriarchal capitalist order. For this neither subversion nor escape is sufficient.

Beyond their disagreement about the nature of oppression and liberation, radical and socialist feminists agree in regarding women's oppression as systemic. Like Adorno, they think the system's oppressive features permeate society as a whole, including the ways in which culture is produced and distributed. Where they differ from Adorno is in their understanding of which class is oppressed, how that oppression operates, and how it could be overcome – and how "the culture industry" figures in all of this.[10] Because of these

---

[9] I use the labels "radical feminism" and "socialist feminism" in the manner proposed by Alison Jaggar, *Feminist Politics and Human Nature* (Totowa, N.J.: Rowman & Allanheld, 1983), and recapitulated by Rosemarie Tong, *Feminist Thought: A More Comprehensive Introduction*, 2d ed. (Boulder, Colo.: Westview Press, 1998).

[10] For an Adornian feminist critique of the "identitarian" logic at work in radical feminism, see Regina Becker-Schmidt, "Critical Theory as a Critique of Society: Theodor W. Adorno's Significance for a Feminist Sociology," in *Adorno, Culture and Feminism*, ed. Maggie O'Neill (London: Sage, 1999), pp. 104–18. Becker-Schmidt's

differences, radical and socialist feminists tend to reject the idea of artistic autonomy that Adorno seems to endorse.

Given both Adorno's proximity to radical and socialist feminist social theory and the risks of rejecting artistic autonomy, one wonders whether elements of Adorno's autonomism are worth salvaging for a feminist cultural politics that is informed by a systemic critique of society as a whole. Perhaps a critical retrieval of insights from Adorno's autonomist critique of the culture industry would allow feminists to address Devereaux's concerns while maintaining legitimate criticisms of artistic autonomy's exclusionary content and consequences.

5.2 THE CULTURE INDUSTRY

Complex Autonomy

A first step in this direction is to show that Adorno's idea of artistic autonomy differs significantly from the idea that radical and socialist feminists reject. His critique of the culture industry in *Dialectic of Enlightenment* uses Marx's dialectic of the commodity to rework eighteenth-century enlightenment aesthetics. Central to Adorno's critique, although barely thematized in the book itself, is the understanding of autonomy captured in Kant's description of the beautiful as purposiveness without purpose (*Zweckmässigkeit ohne Zweck*). Adorno translates this description into a complex conception with wide-ranging social implications. On Adorno's conception, purposiveness without purpose would both require and promote the three interlinked forms of autonomy that I mentioned in Chapter 1: the internal and self-critical independence of authentic artworks, the relative independence of (some of) high culture from the political and economic system, and the autonomy of political and moral agents. I shall label these three forms "internal," "societal," and "personal" autonomy, respectively. Adorno criticizes the culture industry for undermining all three types of autonomy.

As this preliminary description suggests, although Adorno emphasizes artistic autonomy, the version he emphasizes is more

> diagnosis of women's oppression links the hegemony of specific societal spheres (primarily corporate and government sectors) with specific gender hierarchies.

complex and dialectical than the idea many feminists reject. To begin with, Adorno's concept of the artwork's internal autonomy is substantial rather than merely formal. Authentic works of art have an internal dialectic of content and form, he says, and this dialectic expresses the contradictions of society as a whole. Moreover, in carrying out a dialectic of content and form, authentic works challenge their own self-constitution and thereby challenge the society that makes them possible. They force people to confront society's unresolved tensions, and they point toward resolutions that artworks themselves cannot accomplish.

Adorno also links substantial internal autonomy to a societal autonomy that is itself dialectical. On his account, the relative independence some art has achieved in capitalist societies is itself dependent on political and economic developments that he strongly criticizes. So Adorno does not simply celebrate art's societal autonomy, nor does he regard it as ideologically neutral. Although societal autonomy makes possible a crucial mode of social criticism and utopian memory, it also reflects a class-based division of labor that Adorno rejects.

Where he goes wrong, as I have argued elsewhere, is in making internal and societal autonomy a precondition for art's social-critical capacities.[11] In effect, this rules out most forms of folk art, popular art, mass art, and site-specific art as potential locations of social criticism. Such locations include many forms of art making that feminist historians have retrieved and feminist artists have promoted. The other problem with Adorno's dialectical autonomism is that it discounts the entwinement of autonomous art with other social institutions. As many feminists have shown, autonomous art is not as independent from the culture industry and other social institutions as Adorno's language sometimes suggests. Although a distinction can be made out between autonomous art and other forms of cultural

---

[11] See "Models of Mediation," in Zuidervaart, *Adorno's Aesthetic Theory*, pp. 217–47. A shorter version of this chapter was published as "The Social Significance of Autonomous Art: Adorno and Bürger," *Journal of Aesthetics and Art Criticism* 48 (Winter 1990): 61–77. These writings focus on Adorno's notion of truth (*Wahrheitsgehalt*) in art, a notion that includes both social-critical and utopian elements. In the current context, I discuss only the question of social-critical capacity.

production, this distinction is not fixed, nor is it the crucial one for sorting out the social-critical capacities of specific cultural phenomena.

But this argument does not get at the central problems in Adorno's critique of the culture industry. It does not suffice to argue, as I did, that autonomy is not a precondition for art's social-critical capacities. Nor does it suffice to argue, as Deborah Cook does, that some products of the culture industry might be capable of the sort of internal autonomy that Adorno reserved for authentic works of modern art.[12] The problems in Adorno's critique of the culture industry do not arise simply from the way he reserves internal autonomy for certain works of high art. Rather, they arise from how he conceptualizes internal autonomy and from how he connects internal autonomy with both the societal autonomy of art and the personal autonomy of political and moral agents.[13] These connections must be reexamined, along with the idea of autonomy that allows the connections to be posited in the first place. Of particular importance for feminist criticisms of artistic autonomy is the connection between internal and societal autonomy.

Democratic Monopoly

Let us consider, then, how Adorno's critique of the culture industry accounts for art's societal autonomy and implicitly connects this with

---

[12] Deborah Cook, *The Culture Industry Revisited: Theodor W. Adorno on Mass Culture* (Lanham, Md.: Rowman & Littlefield, 1996). Cook suggests that perhaps "some products of the culture industry already follow the model for cultural practice with political import which Adorno discovered in some works of high modern art" (p. 129). Such products would have to achieve "a degree of autonomy" that allows them to "break the stranglehold of reification and ... narcissism, holding out the promise of independent forms of communication between more rationally and instinctually robust individuals" (p. 128). Here Cook is more faithful to Adorno than I consider warranted. For she adopts, without serious challenge, the concept of *internal* autonomy that underlies Adorno's critique.

[13] In "Adorno on Mass Societies," *Journal of Social Philosophy* 32 (Spring 2001): 35–52, Deborah Cook demonstrates that Adorno's general critique of late capitalism links the development of a stratified mass society with the spread of narcissism. This casts doubt on a Habermasian "faith in [the] vitality of civil society or the lifeworld" (p. 39), she argues. There may be an inconsistency here between her embrace of Adorno's general critique and her earlier attempt to rescue some culture-industrial products from Adornian dismissal. See also Deborah Cook, *Adorno, Habermas, and the Search for a Rational Society* (New York: Routledge, 2004).

the internal autonomy of certain artworks. *Dialectic of Enlightenment* lays this out in two passages, which Cook also cites.[14] The first passage argues that the culture industry "proves to be the goal of ... liberalism," both culturally and politically (*DE* 104–6/156–9). The argument compares developments in the United States with ones in Germany. Whereas the United States, with its more democratic culture and polity, led the way toward "the monopoly of culture," the failure of democratic *Kontrolle* to permeate life in Germany exempted leading universities, theaters, and museums from market mechanisms: protection by local and state governments gave them a measure of freedom from commercial domination. At the same time, the market for literature and published music could rely on the purchasing power of homage paid to an unfashionable artistic quality. What really does contemporary artists in, says Adorno, is the pressure to blend into commercial life as "aesthetic experts." They gain freedom from political domination only to become slaves of a "private monopoly of culture" in which the imposed tastes of the defrauded masses keep artists in their place. Prefascist Europe, by contrast, lagged behind the trend toward culture monopoly. This lag left "a remnant of autonomy" for intellectual activity (*Geist*). In other words, whatever societal autonomy the arts enjoyed in prefascist Europe was a function of less-than-democratic political arrangements.

There are two obvious problems with this explanation. First, it does not explain the societal autonomy of the arts in the United States, where the development of democratic forms of polity coincides historically with the institutionalization of autonomous art. Although Adorno quotes Alexis de Tocqueville to negative effect in this passage, he never takes up Tocqueville's discussion of the voluntary associations that spawned intellectual activity in the United States when Germany still relied on state protection and control of culture. Yet one cannot simply ignore the prominence of voluntary associations in American culture, and their relative weakness in Germany, when one explains the social history of autonomous art and its relation to the culture industry. Local music societies, women's literary clubs, and philanthropically funded art museums did not

---

[14] Cook, *The Culture Industry Revisited*, p. 107.

disappear with the rise of mass media in the United States. Instead, they flourished and multiplied, even during the years when Nazis seized control of state-protected cultural organizations in Germany. To describe the American scene as a "culture monopoly," as Adorno does, is to ignore the crucial role played by such noncommercial and nongovernmental organizations.

The other problem with Adorno's account is that it fails to explain why Hitler and his henchmen would attack not only the state-protected institutions of education, scholarship, and art but also the commercial institutions of mass culture. Surely the Nazis found some threat and resistance there, and not simply the "conformism" Adorno associates with the culture industry. Apart from vague appeals to the fact that consumers are not completely duped by the culture industry, Adorno does not explore the critical and subversive tendencies inherent to culture-industrial production, as distinct from particular products of such production. Nor is this surprising, since Adorno's approach largely precludes the "truth potential" of art that does not fit his models of internal and societal autonomy.

As an aside, we may observe that the problem just noted has a subterranean history in Adorno's own response to the Nazi regime. If one can see continuity between Heidegger's ontology of existence from the 1920s and his subsequent pro-Nazi speeches and actions in the 1930s, perhaps one can also detect a certain consistency between the political assumptions in Adorno's critique of the culture industry from the 1940s and some problematic journalistic pieces from the Nazi era. In one, written in Germany in 1933 but wisely never published, Adorno endorses not only state ownership of radio broadcasting but also state-authorized removal of jazz and hit tunes from the airwaves, in favor of more demanding music such as Beethoven's late string quartets. This truly astonishing position – written during the first year of Hitler's regime! – reflects a "political myopia" (Müller-Doohm) not altogether absent from Adorno's subsequent critique of the culture industry.[15] Nor would an appeal to Adorno's mature aesthetics actually deflect the charges of "opportunism" and

---

[15] For a summary and discussion of this and similar essays, see Stefan Müller-Doohm, *Adorno: A Biography*, trans. Rodney Livingstone (Cambridge: Polity Press, 2005), pp. 183–5 and 525–7n42–8.

"hypocrisy" that arose in 1963 when a student discovered a closely related essay published by Adorno in 1934. The essay in question praised a National Socialist composer, and it appeared in a music journal that had become an official mouthpiece of the Nazi youth organization. Although I agree with Espen Hammer that Adorno, like other intellectuals of the time, faced a "deep dilemma" in figuring out how to resist an oppressive regime, much of Adorno's own dilemma stems from his inadequate understanding of the nexus between politics and culture. Unlike Hammer, I would not look to Adorno's mature aesthetics for an "answer to this dilemma."[16]

Total Marketability

The second passage in *Dialectic of Enlightenment* to be considered occurs in a section on the "pseudoindividuality" said to prevail in the culture industry (*DE* 124–31/181–9). Here Adorno argues that the culture industry involves a change in the commodity character of art, such that art's commodity character is deliberately acknowledged and art "abjures its autonomy" (*DE* 127/184). Art's autonomy was always "essentially conditioned by the commodity economy," says Adorno, even when autonomy took the form of negating social utility (*gesellschaftliche Zweckmässigkeit* – literally, "societal purposiveness"). Internally autonomous works – works that "negated the commodity character of society by simply following their own inherent laws" – were still commodities. Adorno claims that their purposelessness was sustained by "the anonymity of the market," whose demands are so diversely mediated that the market partially frees the artist from specific requirements (*DE* 127/184). He acknowledges, of course, that the artist's market-mediated freedom is also a "freedom to starve" (*DE* 104/157) and contains an element of untruth. Yet the proper way for artists to counter this untruth, he says, is neither to deny nor to flaunt the commodity character of art but to assimilate the contradiction between market and autonomy "into the consciousness of their own production" (*DE* 127/185).

The culture industry, by contrast, dispenses entirely with the "purposelessness" that is central to art's autonomy. Once the

[16] See Espen Hammer, *Adorno and the Political* (New York: Routledge, 2005), pp. 50–3.

demand that art be marketable (*Verwertbarkeit*) becomes total, the internal economic structure of cultural commodities shifts.[17] Instead of promising freedom from societally dictated uses, and thereby having a genuine use value that people can enjoy, the product mediated by the culture industry has its use value *replaced* by exchange value: "[E]njoyment is giving way to being there and being in the know, connoisseurship by enhanced prestige. ... Everything is perceived only from the point of view that it can serve as something else. ... Everything has value only in so far as it can be exchanged, not in so far as it is something in itself. For consumers the use value of art, its essence, is a fetish, and the fetish – the social valuation [*gesellschaftliche Schätzung*] which they mistake for the merit [*Rang*] of works of art – becomes its only use value, the only quality they enjoy" (*DE* 128/186). Hence, the culture industry dissolves the "genuine commodity character" that artworks once possessed when exchange value developed use value as its own precondition and did not drag use value along as a "mere appendage" (*DE* 129–30/188).

In this second passage, then, the societal autonomy of art depends on the functioning of a capitalist market prior to the development of monopolistic tendencies. If Adorno were entirely consistent with this

---

[17] Revisions introduced into the text before it was published in 1947 disguise the fact that Adorno and Horkheimer are employing the analysis of the commodity worked out by Karl Marx in *Das Kapital*. Where the 1947 and subsequent editions mention "eine Verschiebung in der inneren ökonomischen Zusammensetzung der Kulturwaren," the 1944 version more straightforwardly points to "eine Verschiebung in der Zusammensetzung der Kulturwaren nach Gebrauchswert und Tauschwert." In the English translation, then, the phrase "a shift in the inner economic composition of cultural commodities" substitutes for what would have been more clearly in line with Marx's analysis, namely, "a shift in the composition of cultural commodities in terms of use value and exchange value." See *DE* 128/185 and the related note on *DE* 271/185. A similar obscuring of Adorno's Marxian categories occurs in the revised version of his essay "On the Fetish-Character in Music and the Regression of Listening." It is this revised version that has been translated into English and widely anthologized. Such revisions have helped shield Adorno's Anglo-American readers from the economic claims that support his critique of the culture industry. Although Richard Leppert does not take up these claims in detail, his superb commentaries on Adorno's life and work, and his positioning of the "Fetish-Character" essay among related writings by Adorno, should help uncover the Marxian roots to Adorno's critique of the culture industry. See *Essays on Music: Theodor W. Adorno*, ed. Richard D. Leppert, trans. Susan H. Gillespie et al. (Berkeley: University of California Press, 2002). The "Fetish-Character" essay, in a translation modified by Leppert, occurs on pp. 288–317.

analysis, he should have concluded that economic preconditions for the societal autonomy of art had disappeared by the time he wrote his critique of the culture industry, and long before he began writing his *Aesthetic Theory*. But that conclusion would contradict the way in which he pits modern art against the culture industry, and it would make his critique of the culture industry reactionary and inconsistent with the progressive impulses of his social theory.[18]

Nevertheless, Adorno's critique has made an innovative and productive move, namely, to interpret Kant's notion of autonomy through an updated reading of Marx's dialectic of the commodity, such that questions of internal and societal autonomy become inextricably linked. Too many of Adorno's Anglo-American critics, in their desire to secure legitimacy for "mass art" or "mass culture," overlook this move. They prefer instead to argue that Adorno misunderstood mass art because he misconstrued "Kant's analysis of free beauty as a theory of art,"[19] or that Adorno did not understand genre-specific characteristics, such as the importance of individual performance in jazz and the salience of collaborative recordings in rock.[20] In effect, these critics suggest that Adorno did not really understand American "popular culture" and illicitly measured it by

---

[18] Deborah Cook's *The Culture Industry Revisited* notes the problem here, but she does not seem to feel its full weight. She finds three "extrinsic conditions" for modern art's autonomy in Adorno's writings: (1) a cultural lag behind the culture industry, (2) the advancement of artistic techniques, and (3) the psychological dispositions of the artist. Of these, only the first is mildly economic, and it is hardly reassuring with regard to prospects for the societal autonomy of high art, much less the societal autonomy of popular culture and of mass culture. If one wants to argue as Cook does for the internal autonomy of artworks and of certain products of the culture industry, one must either abandon the notion that this internal autonomy is somehow linked to economic conditions or provide a different account of those economic conditions than Adorno himself provides. The first option requires one to give up Adorno's dialectical interpolations of culture and economy. The second option, although less problematic from a Critical Theory perspective, is also more difficult.

[19] Noël Carroll, *A Philosophy of Mass Art* (Oxford: Clarendon, 1998), p. 105. It is curious that Carroll does not discuss Kant's own attempt to distinguish between "fine art" and "agreeable arts" where, it seems to me, Kant himself turns his analysis of free beauty into a theory of art. See §§43–46 in Immanuel Kant, *Critique of the Power of Judgment*, ed. Paul Guyer, trans. Paul Guyer and Eric Matthews (Cambridge: Cambridge University Press, 2000), pp. 182–7.

[20] Theodore Gracyk, *Rhythm and Noise: An Aesthetics of Rock* (London: I. B. Tauris, 1996), pp. 149–73.

the standards of European bourgeois art. Such criticisms, while interesting in their own right, sidestep the challenge Adorno's idea of autonomy poses for politically inflected theories of contemporary culture.[21] The criticisms do not consider how culture and economy intersect, or what this intersection means for emancipatory politics. Feminists, however, cannot afford to ignore Adorno's challenge.

5.3 CULTURE, POLITICS, AND ECONOMY

Earlier I mentioned Mary Devereaux's concern that feminist criticisms of artistic autonomy might undermine the "protected space" required for feminist cultural politics. By itself, Adorno's critique of the culture industry would not help resolve this dilemma. His critique includes a concept of internal autonomy not unlike the notion of artistic autonomy that many feminists reject. Yet feminists need to take up the larger challenge of Adorno's complex idea of autonomy, namely, to develop a critical understanding of the systemic roles fulfilled by culture in contemporary society.

Spaces of Resistance

A good place to begin is to revisit Adorno's claims concerning the internal economic structure of cultural commodities. On the one hand, his analysis illuminates the essential role of hypercommercialization in contemporary Western societies. Whether homespun or esoteric, whether radical or timid, no cultural good is protected from the juggernaut of consumer capitalism and the hegemony it secures. On the other hand, Adorno's analysis has an unfortunate consequence: it portrays many cultural goods as *no more than* hypercommodities whose exchange value has replaced their use value. This portrait is unfortunate because it inadvertently endorses the tendency toward hypercommercialization that Adorno opposes. Adorno fails to acknowledge sufficiently that taking pleasure in

---

[21] The same avoidance occurs in Patrick Brantlinger's highly influential interpretation of Adorno's critique as an example of "negative classicism," in *Bread and Circuses: Theories of Mass Culture as Social Decay* (Ithaca, N.Y.: Cornell University Press, 1983).

exchange value is not the sole function of culture-industrial transactions – despite the "art is business" mentality of production and distribution companies, and despite the "I know what I like" self-understanding of culture "consumers."[22]

Of course, one can hardly deny that, under consumer capitalist conditions, cultural producers sell exchange values as use values and consumers buy them to be hip and fashionable. Simultaneously, however, and unavoidably, they also engage in cultural practices mediated by cultural goods. The "value" of these practices and goods cannot be subsumed under the economic categories of use value and exchange value. Accordingly, contemporary feminists must ask how hypercommercialization either enhances or undermines such cultural practices, in which respects, and to what effect. One's answer would indicate what, beyond offering a critique of consumer capitalism, would be an appropriate cultural politics.

Because these matters are difficult to sort out, I do not pretend to give satisfactory answers in a single chapter. But let me outline an approach that could prove fruitful for feminist cultural politics. Russell Keat has argued that cultural activities such as broadcasting, the arts, and academic research, and "the institutions within which they are conducted, should be 'protected' in various ways from the operation of the market."[23] Keat's argument is an economic counterpart to Mary Devereaux's political argument that the arts need a "protected space," by which she means, primarily, a space protected from censorship and other forms of "political interference." While acknowledging the point to both arguments, I would add that adequate political protection requires adequate economic protection, and vice versa. More important, I would also claim that protection is not enough: structural alternatives are required to the dominant economy and the dominant political system.

Why do such cultural activities need protection from both "political interference" and the "operation of the market"? The reason is

[22] Together with several colleagues in communications, film studies, history, and musicology, I explore other functions of culture-industrial transactions in *Dancing in the Dark: Youth, Popular Culture, and the Electronic Media*, ed. Roy Anker (Grand Rapids, Mich.: Eerdmans, 1991).
[23] Russell Keat, *Cultural Goods and the Limits of the Market* (London: Macmillan, 2000), p. ix.

that in contemporary Western societies a legitimate differentiation of cultural and societal spheres has occurred in tandem with an illegitimate colonization of ordinary life and cultural institutions by both corporate and governmental systems.[24] These systems mutually reinforce each other's dominance and, in doing so, they reinforce patriarchal patterns in culture. That is why political protection without economic protection, or vice versa, usually proves inadequate. Moreover, mere protection, accompanied perhaps by appeals to the intrinsic worth of education or research or the arts or entertainment, is not sufficient to resist pressures toward hypercommercialization and performance fetishism. Certainly, to thrive, such cultural activities need to have economically and politically "protected spaces." If the spaces are themselves created by the corporate economy or state agencies or are overly dependent on these, however, then the activities conducted within them remain subservient to corporate or state dictates and are ruled by the systemic logics of money and power.

Hence, as radical feminists have seen more clearly than most, countereconomic and counterpolitical spaces need to emerge from cultural activities and cultural agents themselves. That means setting up and strengthening organizations, media, and social networks whose economy is noncommercial and whose public voice is not state-sanctioned. Such spaces are not simply a "counterculture" that

---

[24] The main lines of this answer stem from Habermas's diagnosis of modernization in *The Theory of Communicative Action*. I explore the relevance of this diagnosis for the arts in "Postmodern Arts and the Birth of a Democratic Culture," in *The Arts, Community and Cultural Democracy*, ed. Lambert Zuidervaart and Henry Luttikhuizen (New York: St. Martin's, 2000), pp. 15–39, where I introduce the concepts of hypercommercialization and performance fetishism. By "hypercommercialization" I mean a tendency to make commercial potential the primary or sole reason for building and maintaining cultural organizations, as seen, for example, in moves to "privatize" state-funded schooling and to market the economic benefits of arts organizations. This tendency comes with a full-scale celebration of exchange value that makes it increasingly difficult to raise and address questions of cultural need and cultural norms. By "performance fetishism" I mean a tendency toward internal bureaucratization of cultural organizations in response to government demands for administratively manageable certification of competence and output, with the result that strategic calculations take precedence over the cultural missions of such organizations. Examples include moves to "retool" school curricula and pedagogies for the sake of competitiveness in the global economy and to "rationalize" arts organizations in accord with cost-benefit criteria.

would be noneconomic or apolitical. A countereconomy is definitely an economy, but it operates on noncommercial principles, whether nonprofit, cooperative, or communal. So too, a counterpolitics is definitely political, but it is not simply caught up in lobbying, party politics, and occupying legislative, administrative, or judicial positions of power. Even if the long-term goal were to break the dominance of corporate and state systems, as it seems to be in the antiglobalization movement, fostering countereconomic and counterpolitical spaces would still be internally important and externally necessary. Internally important, because otherwise the colonization of culture faces little resistance. Externally necessary, because there is hardly any other way to mobilize economic and political opposition to systems wielding enormous clout, with oppressive results.

Imaginative Cogency

As Adorno's linking of internal and societal autonomy suggests, countereconomic and counterpolitical spaces must foster intrinsically worthwhile cultural practices. This applies to the arts just as much as it applies to education and research. Perhaps a suitably expanded concept of "aesthetic value"[25] would provide a constructive alternative to Adorno's analysis and help resolve the dilemma within feminism concerning artistic autonomy. By "suitably expanded" I mean two things. First, I mean a concept that does not restrict aesthetic merit to formal qualities of discrete art objects but encompasses the qualities of the practices in which people engage when they create and experience artistic products and events. Second, I mean a concept of aesthetic merit that does not restrict the relevant practices to those typical of high art but ranges over popular art, mass art, and site-specific art as well. In Chapter 1 I describe this concept as "imaginative cogency." Imaginative cogency is a horizon

---

[25] Here I take a cue from Peggy Zeglin Brand, "Revising the Aesthetic-Nonaesthetic Distinction: The Aesthetic Value of Activist Art," in *Feminism and Tradition in Aesthetics*, ed. Peggy Zeglin Brand and Carolyn Korsmeyer (University Park: Pennsylvania State University Press, 1995), pp. 245–72. Brand argues that philosophical aesthetics needs to move away from "the rigidity of the traditional aesthetic-nonaesthetic distinction and toward a revised notion ... of aesthetic value" (p. 260). My proposal differs from Brand's in emphasizing cultural practices rather than artworks.

of expectations governing the intersubjective exploration, presentation, and interpretation of aesthetic signs. Within this horizon, art products and art events are expected to elicit and sustain open-ended exploration, to present multiple and unexpected nuances of meaning, and to lend themselves to creative interpretation. And these imaginative processes are expected to occur with degrees of complexity, depth, and intensity that are both appropriate to the particular products and events and intrinsically worthwhile.

The concept of imaginative cogency helps articulate a notion of autonomy that neither privileges authentic works of modern art nor construes artistic autonomy as a last bastion of social critique, à la Adorno. But it also resists either reducing aesthetic merit to the outcome of struggles for power or treating artistic autonomy as a merely strategic concern, in the manner of some feminists. As was indicated in Chapter 1, the central normative question concerning autonomy would be which contemporary art practices are better able to foster creative and critical dialogue. Accordingly, concerns about artistic autonomy would shift from artworks and culture-industrial products as such to the quality of the practices undertaken by cultural workers and cultural publics. There is nothing in this notion of artistic autonomy to preclude its application to popular, mass, and site-specific art. Yet it retains a critical edge and does not simply become a classificatory category.[26] In fact, it derives in large part from collaborative and interventionist practices pioneered by feminist artists such as Judy Chicago and Judith Baca.[27]

Given this critical hermeneutic notion of artistic autonomy, one can also ask which institutional patterns and societal structures are more likely to foster the sorts of practices in which imaginative

---

[26] This approach has affinities with Miriam Hansen's attempt to think through Adorno's writings on film and mass culture in "Mass Culture as Hieroglyphic Writing: Adorno, Derrida, Kracauer," *New German Critique*, no. 56 (Spring–Summer 1992): 43–73. Hansen tries to mobilize "the split between mass-cultural script and modernist *écriture*" into "a stereoscopic vision that spans the extremes of contemporary media culture," as both serving "the ever more effective simulation of presence" and birthing "a postmodern culture of difference" (p. 73). But her essay does not address the connection of this vision to emancipatory politics.

[27] See Lambert Zuidervaart, "Creative Border Crossing in New Public Culture," in *Literature and the Renewal of the Public Sphere*, ed. Susan VanZanten Gallagher and Mark D. Walhout (New York: St. Martin's, 2000), pp. 206–24.

cogency can be pursued. This question can be made more specific to feminist struggles for justice and recognition that include the creation and experience of art.[28] Because such questions are partly empirical, one cannot settle them in theory, no more than in theory Adorno could legitimately declare all products of the culture industry worthless, and no more than in theory a feminist can legitimately declare all aesthetic criteria to be simply ideological. Yet there is one respect in which a social-theoretical answer needs to be attempted. This has to do with the relationships among economy, polity, and culture that Adorno makes central to his own critique of the culture industry.

New Public Spheres

In the two passages I have discussed, Adorno depicts the societal autonomy of the arts as the function, historically, of a paternalist state and premonopolistic capitalism. Although understandable both as a reflection of Adorno's own European experience and as an articulation of his background theory of state capitalism, such an account of societal autonomy misses a crucial feature of cultural politics and cultural economics in the United States. I mentioned earlier his failure to take up Tocqueville's discussion of voluntary associations in America. This is doubly unfortunate. Not only did these types of organizations resist the grid of capitalism in either its entrepreneurial or its monopolistic stages, but they also have provided alternative sites for the development of cultural practices outside the strict confines of church, state, and other institutions of control. Freestanding schools, music societies, libraries, and literary clubs, many of them founded or led by women, may have had as much impact as the capitalist market did on the development of culture in the United States. This is not to suggest that cultural organizations have a more decisive role in the development of culture than economic forces have. Nor is it to suggest that such sites were

---

[28] The relationship between struggles for economic justice and struggles for cultural recognition is especially crucial for the emancipation of what Nancy Fraser calls "bivalent collectivities," such as oppressed groups constituted along the lines of either race or gender. See *Justice Interruptus: Critical Reflections on the "Postsocialist" Condition* (New York: Routledge, 1997), pp. 11–39.

completely independent from class interests, patriarchal patterns, and a capitalist economy. Rather, the type of economy that helped give rise to twentieth-century American culture cannot be captured in either a simple market model or the model of state capitalism that informs *Dialectic of Enlightenment*. It is a three-sector economy, and it includes a voluntary component that mediates the development of cultural practices.[29]

If my hypotheses are on the right track, then the account to be given of the societal autonomy of art must go beyond Adorno's theses about the political backwardness of prefascist Europe and the commodity character of artistic products. It needs to incorporate the role of civil society and public spheres, not only historically but also today. That, ironically, is the missing link both in the Kantian account of "fine art" (*schöne Kunst*) and in Adorno's Marx-inflected usage of Kantian notions to critique the culture industry – ironically, because, if Habermas's *The Structural Transformation of the Public Sphere* is right, the rise of fine art as an autonomous social institution in eighteenth-century Europe is itself intimately entwined with the development of a relatively independent bourgeois public sphere.[30]

As historians such as Joan Landes have shown,[31] contemporary feminism must consider whether such institutionalizations are irreversibly bound to the masculinist structures in which they first

---

[29] The three sectors of the U.S. economy are the corporate, government, and "independent" or "third" sectors. I have discussed the shortcomings of standard economic theories of the third sector in a paper titled "Short Circuits and Market Failure: Theories of the Civic Sector," presented at the Twentieth World Congress of Philosophy in Boston (August 1998), http://www.bu.edu/wcp/Papers/Soci/SociZuid.htm.

[30] Jürgen Habermas, *The Structural Transformation of the Public Sphere: An Inquiry into a Category of Bourgeois Society* (1962), trans. Thomas Burger and Frederick Lawrence (Cambridge, Mass.: MIT Press, 1989). For a brief summary, see Jürgen Habermas, "The Public Sphere: An Encyclopedia Article" (1964), trans. Sara Lennox and Frank Lennox, *New German Critique* 1 (Fall 1974): 49–55. Habermas revisits these topics in "Concluding Remarks" and "Further Reflections on the Public Sphere," in *Habermas and the Public Sphere*, ed. Craig Calhoun (Cambridge, Mass.: MIT Press, 1992), pp. 462–79 and 421–61, respectively. See also the chapter "Civil Society and the Political Public Sphere" in Habermas's *Between Facts and Norms: Contributions to a Discourse Theory of Law and Democracy*, trans. William Rehg (Cambridge, Mass.: MIT Press, 1996), pp. 329–87.

[31] See especially Joan B. Landes, *Women and the Public Sphere in the Age of the French Revolution* (Ithaca: Cornell University Press, 1988).

emerged, or whether, through concerted counterhegemonic effort, they have become and can become sites of opposition and transformation. For such effort to succeed, countereconomic forces are required. Habermas himself does not consider the alternative economic underpinnings of some public spheres. In this he remains too close to Adorno's critique of the culture industry.[32] That has not escaped the attention of feminists who incorporate Habermas's emphasis on public spheres but who see that a feminist cultural politics might require countereconomic structures. Rita Felski, for example, partially recognizes the need for alternative economic support to a "feminist counter-public sphere." Yet she remains ambivalent about the prospects for developing such support, and she envisions the feminist counterpublic sphere as operating through "a series of cultural strategies" both internal and external to "existing institutional structures."[33] This raises a problem, it seems to me. Existing institutional structures such as "the educational system" are just as vulnerable to hypercommercialization and performance fetishism as are cultural strategies external to them. While I understand why Felski rejects Adorno's supposed "privileging of a modernist aesthetic as a site of freedom" and questions his apparent "diagnosis of the modern world as a totally administered society with no possibility of genuine opposition or dissent,"[34] I do not see how a

---

[32] Part of the problem is that Adorno's theses tend to be monocausal. It is always tricky to sort out causes from effects in social-theoretical and historical explanations, and even more tricky to include enough of the relevant factors. Did the arts achieve societal autonomy because market forces replaced court and church patronage? Or did market forces replace patronage in the arts because participants in the arts were seeking societal autonomy? Did the arts in prefascist Europe retain a measure of autonomy because they enjoyed state protection from monopolistic market forces? Or did the state protect the arts in Europe in order to retain societal dominance against monopolistic market forces? All of these lines of inquiry are inadequate, since they assume a single direction of causality and leave out many other potentially relevant factors. At the same time, of course, theoretical simplification and abstraction can yield an account rich in explanatory potential.

[33] Rita Felski, *Beyond Feminist Aesthetics: Feminist Literature and Social Change* (Cambridge, Mass.: Harvard University Press, 1989), p. 171.

[34] Ibid., p. 163. For a shorter account of this attempt to go "beyond feminist aesthetics," see Rita Felski, "Why Feminism Doesn't Need an Aesthetic (And Why It Can't Ignore Aesthetics)," in Brand and Korsmeyer, *Feminism and Tradition in Aesthetics*, pp. 431–45.

feminist counterpublic sphere would escape Adorno's diagnosis if it did not actually have countereconomic support.

So, although indebted to the pioneering work of both Habermas and Felski, my response to Adorno's critique of the culture industry takes a different tack. I argue against Adorno that the societal autonomy of art depends on a number of interrelated social factors. While structural shifts in the dominant economy and polity have significant implications for the autonomy of art, they are linked to noncommercial and nongovernmental developments that also inform structural shifts in the dominant economy and polity. The twentieth-century shift from monopoly capitalism to consumer capitalism, for example, may well go hand in hand with the development of new public spheres where struggles for recognition and justice exceed the boundaries of state-sanctioned discourse, and with a proliferation of cultural organizations whose third-sector economy need not turn products into hypercommodities.

An account along these lines would not assume that "purposiveness without purpose" was definitive of fine art in Kant's own day or that the current prospects for autonomous art depend primarily upon aesthetic qualities internal to the work of art. Instead, the account would make more of the "communicability" and "sociability" that Kant links with aesthetic judgment.[35] It would also explore the matrix of civil society and third-sector economy that gives birth to diverse forms of cultural creation, not only in North America but also around the world.

Accordingly, my own alternative to Adorno's idea of autonomy involves three steps. First, I replace the notion of a work's internal autonomy with a notion of the autonomy of certain cultural practices. Next, I revise Adorno's political and economic account of art's societal autonomy. Finally, in other publications, I indicate how certain practices, as made possible by certain institutions, can foster a more dialogical type of personal autonomy characterized by creative co-responsibility.[36] This account would have three outcomes. First, the question whether culture-industrial products can have emancipatory

---

[35] See Kant, *Critique of the Power of Judgment*, §§39–41, pp. 171–8.
[36] I elaborate this alternative and explore its implications for public policy in a book manuscript provisionally titled *Art in Public: Politics, Economics, and a Democratic Culture*.

potential would turn into another question, one that asks whether cultural organizations can be fashioned where cultural products of many sorts can be taken up in autonomous cultural practices – organizations that, by virtue of their political and economic positioning, resist systemic colonization. Second, the blunt causal assumptions of Adorno's critique of the culture industry would give way to a more textured diagnostic model. And, third, the conflict in feminism between strategically appealing to artistic autonomy and theoretically rejecting it would dissolve. It would dissolve into a reconfigured idea of autonomy. Reconfigured, autonomy would no longer be the preserve of discrete products, branches, or agents of culture. Rather, it would be a multidimensional condition to be fashioned and won, ever anew, in struggles for recognition and justice.[37] That is why, as Adorno says, "this is not a time for political works of art" – not because "politics has migrated into the autonomous work of art," however, but because "autonomy" has migrated beyond the boundaries of Adorno's paradoxically modernist aesthetics.

[37] Perhaps this idea of autonomy would be a social-philosophical counterpart to the attempt "to explore the possibilities of a post-structuralist aesthetics (hence, a postanalytic and a postfeminist aesthetics)" in Joseph Margolis, "Reconciling Analytic and Feminist Philosophy and Aesthetics," in Brand and Korsmeyer, *Feminism and Tradition in Aesthetics*, pp. 416–30. Margolis's essay responds to Joann B. Waugh's "Analytic Aesthetics and Feminist Aesthetics: Neither/Nor?" in the same volume, pp. 399–415.

6

Ethical Turns

> Wrong life cannot be lived rightly.
>
> Adorno, *Minima Moralia*[1]

Student activists in the 1960s who had absorbed Adorno's critique of "the administered world" became impatient with his apparent lack of political alternatives. They asked, in effect, "What is to be done?" According to Adorno's social philosophy, however, V. I. Lenin's famous question can no longer be posed in the same way. That is one

---

Portions of this chapter were developed and presented in my graduate seminar "The Global and the Local: Capitalism, Culture, and Democracy." I wish to thank the participants for the stimulating conversations we enjoyed.

[1] *MM* §18, p. 39/43. The sentence in German reads: "Es gibt kein richtiges Leben im falschen." Jephcott's translation retains the pointedness of Adorno's aphorism but changes its point. A literal but less literary rendering would be "There is no good life in a false one" or "There is no good life within that which is false," or, in Livingstone's translation, "There can be no good life within the bad one"; see Theodor W. Adorno, *Problems of Moral Philosophy*, ed. Thomas Schröder, trans. Rodney Livingstone (Stanford: Stanford University Press, 2000), p. 1. "Richtiges Leben" alludes to the primary topic of *Minima Moralia*, whose title parodies the Aristotelian *Magna Moralia*. Adorno describes the subject matter of his "melancholy science" as "the teaching of the good life" (*die Lehre vom richtigen Leben*) (*MM*, p. 15/13). The phrase "melancholy science" (*traurige Wissenschaft*), in turn, inverts the title of Nietzsche's book *The Gay Science* (*Fröhliche Wissenschaft*). For reasons to translate Adorno's *richtiges Leben* as "good life" rather than "right living," see the translator's note to Gerhard Schweppenhäuser, "Adorno's Negative Moral Philosophy," in *The Cambridge Companion to Adorno*, ed. Tom Huhn (Cambridge: Cambridge University Press, 2004), p. 349n1. On the title and topics of Adorno's book, see Raymond Geuss, *Outside Ethics* (Princeton: Princeton University Press, 2005), pp. 235–7.

of the sobering lessons to be retained by a social philosophy after Adorno. Yet the question will not disappear so long as one thinks that society as a whole needs to be transformed.

When Adorno said that wrong life cannot be lived rightly, he did not mean that relatively good actions and dispositions are impossible. His point, instead, was that no individual is immune from the corrupting power of a "false society." Further, any attempt to reflect critically on contemporary political or moral prospects must take into account the societal mediation of each individual's life. Adorno's own "reflections from damaged life" begin with the following self-admonition: "[One] who wishes to know [*erfahren*] the truth about life in its immediacy must scrutinize its estranged form, the objective powers that determine individual existence even in its most hidden recesses" (*MM*, p. 15/13). Given such societal mediation, which Adorno links with the consumptive pressures of late capitalist production,[2] even the once revolutionary question "What is to be done?" can take on ideological functions. As he would put it when he criticized student "actionism" (*Aktionismus*) in the 1960s, "[I]f praxis obscures its own present impossibility with the opiate of collectivity, it becomes in its turn ideology. There is a sure sign of this: the question 'what is to be done?' as an automatic reflex to every critical thought before it is fully expressed, let alone comprehended. Nowhere is the obscurantism of the latest hostility to theory so flagrant."[3]

Paradoxically, sensitivity to such ideological potential, combined with the practical disappointments of the New Left, has directed anti-Habermasian critical theorists away from the societal to the ethical. Increasingly, the reception of Adorno's radical social critique has slipped into an ethical turn. There are notable exceptions, of course. Fredric Jameson, for example, heroically christened Adorno's unusual Marxism "a dialectical model for the 1990s."[4] Nevertheless, many attempts to reclaim Adorno from the Habermasians, when not

---

[2] "What the philosophers once knew as life has become the sphere of private existence and now of mere consumption, dragged along as an appendage of the process of material production, without autonomy or substance of its own." *MM*, p. 15/13.

[3] Adorno, "Marginalia to Theory and Praxis," *CM* 276/779. Similar claims, less polemically expressed, occur in the first lecture (May 7, 1963) of Adorno's *Problems of Moral Philosophy* (*PMP* 1–11/9–23).

[4] Fredric Jameson, *Late Marxism: Adorno; or, The Persistence of the Dialectic* (London: Verso, 1990), p. 251.

defending and elaborating his aesthetics, turn his social philosophy into an ethics of "damaged life." The retrieval of Adorno's insights threatens to become a jargon of ethicality.

In fact, one could view the reception of Adorno's social philosophy among subsequent critical theorists as a series of ethical turns. First there occurs a turn toward discourse ethics, led by Habermas, Wellmer, and their colleagues and students. Then anti-Habermasians such as Drucilla Cornell and J. M. Bernstein return to Adorno for an emphatic ethics of the nonidentical. Simultaneously, and with varying degrees of sympathy for both Habermasian discourse ethics and Adornian emphatic ethics, authors such as Seyla Benhabib and Axel Honneth attempt to retrieve those elements of a Hegelian social ethics which seem to have gone missing in both Adorno and Habermas. As will become apparent, on matters ethical and political, my own critical retrieval of Adorno's social philosophy aligns more closely with this third ethical turn than with the other two.

This chapter, which also serves as a conclusion, examines the nexus of politics and ethics in Adorno's thought. I begin by taking issue with versions of emphatic ethics that overlook both the societal reach and the inherent limitations of Adorno's politics. Then, I propose a political alternative that recaptures the social ethics suppressed by emphatic ethics. This proposal leaves behind problematic elements in Adorno's social philosophy but does not endorse Habermas's discourse ethics. I conclude by arguing that, to do justice to Adorno's new categorical imperative, a post-Adornian democratic politics of global transformation is required.

6.1 ADORNO'S POLITICS

Recent reconsiderations of Adorno's politics tend to turn it into an apolitical ethics. This occurs in one of two ways, either by valorizing Adorno's insistence on the autonomy of authentic art and critical thought or by celebrating a stance of enlightened individual resistance. Whereas the first reduces Adorno's politics to an ethics of high culture, the second reduces it to an ethics of personal integrity. Russell Berman and Espen Hammer, for example, argue correctly that Adorno was not the apolitical aesthete that his left-wing critics once made him out to be. What these authors mean by Adorno's "politics," however, comes to little more than maintaining a stance of cultural or personal integrity. Such a reduction is not surprising, given the links among internal, societal, and personal autonomy discussed in previous chapters. Yet the reduction of Adorno's politics to a type of cultural or personal ethics misses some crucial insights in Adorno's social philosophy, and it fails to address the political problems in his aesthetics.

Berman locates the core of Adorno's politics in his lifelong insistence, against pressures from both the right and the left, on the autonomy "of art, of theory, and the individual."[5] I think Berman's interpretation is consistent with Adorno's reluctance to urge political courses of action other than "democratic pedagogy," even when he identified social structures that engender fascist tendencies.[6] Yet it is hard to see how such insistence on cultural autonomy would have much purchase on the state or in political public spheres. If Adorno's politics comes down to urging and enacting a high-cultural change in consciousness, this would not be negligible in itself. Yet it would be quite limited in scope, and inadequate too, given both the complexity of contemporary societies and the radicalness of Adorno's own critique.

The tendency to reduce politics to ethics is even more pronounced in Hammer's *Adorno and the Political*. Hammer says Adorno's politics consists in "ethically informed, micro-interruptive operations" that are "models of responsible exercise of autonomy" in purportedly democratic societies. Acknowledging that Adorno endorses the temporary migration of politics into theory and art, Hammer asks whether such an "ethics of resistance" can be conceived in a coherent fashion.[7] He concludes that Adorno's "strategically elitist stress on disruption from culturally privileged standpoints" is unhelpful for

---

[5] Russell Berman, "Adorno's Politics," in *Adorno: A Critical Reader*, ed. Nigel Gibson and Andrew Rubin (Oxford: Blackwell, 2002), p. 123.

[6] A particularly striking example of this occurs in the essay "Education after Auschwitz," *CM* 191–204/674–90, where Adorno states: "Since the possibility of changing the objective – namely societal and political – conditions [that incubate such events] is extremely limited today, attempts to work against the repetition of Auschwitz are necessarily restricted to the subjective dimension. By this I also mean essentially the psychology of people who do such things" (*CM* 192/675–6). Berman comments on this essay in "Adorno's Politics," pp. 124–6.

[7] Espen Hammer, *Adorno and the Political* (New York: Routledge, 2005), pp. 8, 25.

addressing "any *collective* political project." Nevertheless, Hammer praises the Adornian ethic as an effective counterweight to liberal and Habermasian attempts to restrict politics to "the management of social positivity" and "consensually enforced administration."[8]

By taking this approach, Hammer avoids a number of political questions. To say that Adorno's elitism is merely "strategic" is to avoid asking whether it is politically legitimate. To say that it is unhelpful for *collective* political projects is problematically to presuppose, à la Stanley Cavell's perfectionism,[9] that some genuinely political projects are *not* collective. To pit Adorno's "negativism" against liberal and Habermasian "positivism" is to ignore normative issues concerning the state and political public spheres. All of these potential objections point in the same direction: Hammer has turned Adorno's politics into a type of personal ethics. When, contra Jameson, he declares Marxism not essential to Adorno's political thought, Hammer states: "At the core of [Adorno's] vision of politics lies ... an ethics of resistance – a readiness to think and act such that the space of the political is liberated from the grasp of identity."[10] This summary of Hammer's interpretation encapsulates its problems. For it is unclear what liberating the political from the grasp of identity could possibly mean. It is also doubtful whether Adorno, who opposed the illegitimate imposition of identity, not identity as such, would recognize this as his vision.

Hammer's formulations raise a number of issues for a critical retrieval of Adorno's insights, not least of which is the relation between ethics and politics. While Hammer is surely right that isolating the ethical from the political distorts "what politics can be,"[11] it is also a mistake, both in theory and in practice, to reduce the political to the ethical. Although Adorno does not make this mistake,

---

[8] Ibid., pp. 178–80.
[9] See ibid., pp. 163–6, where a favorable comparison is made between Cavell and Adorno. Hammer says that both Cavell and Adorno regard "conformism and prejudice" as "the supreme threats to democracy," and that both oppose "any account of universality that presupposes an impersonal and pre-given ... structure that unites the community and enables the philosopher to speak representatively" (p. 165). My own discussion in the preceding chapters suggests that to apply these descriptions to Adorno would require so many qualifications as to render them inapplicable.
[10] Ibid., p. 179.   [11] Ibid., p. 158.

Hammer does. Consequently he misses both crucial potential and intractable problems in Adorno's politics.

To begin, one needs to distinguish Adorno's actual political interventions both from compatible political orientations and from his theory concerning the political domain. Hammer tends to elide these distinctions, stating first that "Adorno never developed anything like a theory of politics" and then claiming that "[t]he validity of Adorno's approach to politics cannot be separated from the 'success' of each of his critical interventions."[12] There is something odd about this elision. As Hammer recognizes, Adorno holds that under current conditions a critical theory is itself the best form of political practice. If this theory has little to say about "representative government, international relations, legitimate forms of dissent, and so on,"[13] or if what it says is demonstrably inadequate, then on Adorno's own terms that would pose a *political* problem, precisely by being a *theoretical* deficit. Adorno's critical interventions might well be admirable – I think that they are – but at the same time reflect a deficient understanding of the political domain. If they do, then to celebrate his interventions as exemplars for an "ethics of resistance" would be to turn a political problem into an apolitical solution.

Unlike Hammer, I take Marxism to be central to Adorno's politics in all three senses: political theory, political vision, and actual interventions. Unlike Jameson, however, I think the centrality of Marxism generates oversights as well as insights. It is well known that the Marxist tradition, especially as it came to Adorno via Lukács, is weak in the area of political theory. For significant strands of Marxism, the state and political public spheres are not legitimate in their own right. Rather they are instruments of class struggle whose necessity would largely disappear under postcapitalist conditions. This stance ignores normative issues concerning public justice under current conditions. Friedrich Pollock's theory of state capitalism, to which Adorno partially subscribed, simply exacerbates a strong tendency in the Marxist tradition.

While understandable in the context of Adorno's Marxism, his failure to develop a political theory in the standard sense is a problem, and what he did develop is a poor substitute. That is one reason why my attempts in previous chapters to rescue Adorno's insights

---

[12] Ibid., pp. 1, 8.   [13] Ibid., p. 1.

have not ignored legitimate Habermasian criticisms. Accordingly, the reexamination of Adorno's account of artistic autonomy, begun in Chapter 1, takes on larger political significance. Like Berman and Hammer, I think that Adorno's insistence on autonomy – artistic, theoretical, and personal autonomy – lies at the heart of his politics. Moreover, I have defended this insistence against both Derridean deobjectification (Chapter 1) and feminist dismissal (Chapter 5). Yet I find Adorno's account of autonomy more troublesome than Berman and Hammer do. By pitting the individual artwork or theorist or moral agent against society as a whole, Adorno overlooks the collective practices and institutions without which no authentic artwork or critical theory or moral stance is possible. As Axel Honneth puts it, "Adorno did not make room in his social theory for an autonomous sphere of cultural action in which members of a social group bring their everyday experiences and interests into agreement."[14] Adorno's account of autonomy lacks actual collectivity, even as it presupposes a strong normativity that it cannot articulate. This is so both because Adorno does not allow for noneconomic filters between the economy and culture and because he restricts the psychological mechanisms of healthy ego formation to the internalization of paternal authority.[15] Such an approach holds limited promise for either theorizing or strategizing with respect to political public spheres, which are inherently collective, and which make little sense apart from articulable norms of both process and content.

The standard defense of Adorno at this point is to recall the radicalness of his social critique: it is so radical that it must question all existing forms of collectivity and normativity. This cannot be correct, however, for Adorno does not in fact question all existing forms of collectivity and normativity. Nor can the project of radical critique get off the ground if all such forms are considered ineffective or denied. This is a central insight in Hegel's *Philosophy of Right*.[16]

---

[14] Axel Honneth, *The Critique of Power: Reflective Stages in a Critical Social Theory*, trans. Kenneth Baynes (Cambridge, Mass.: MIT Press, 1991), p. 80.
[15] Here Honneth relies heavily on the path-breaking work of Jessica Benjamin. See Honneth, *The Critique of Power*, pp. 81–92, and Jessica Benjamin, "The End of Internalization: Adorno's Social Psychology," *Telos*, no. 32 (Summer 1977): 42–64.
[16] G. W. F. Hegel, *Elements of the Philosophy of Right*, ed. Allen W. Wood, trans. H. B. Nisbet (Cambridge: Cambridge University Press, 1991).

Indeed, to take up the topics of collectivity and normativity requires a return to aspects of Hegelian political thought that Adorno, like Marx before him, finds problematic. Nor would this be unwarranted, insofar as Adorno is more Hegelian than either his Habermasian or anti-Habermasian successors in Critical Theory: his social critique neither excises the idea of reconciliation nor gives up the category of totality. Adorno's lecture "Wozu noch Philosophie" puts it like this: "Praxis, whose purpose is to produce a ... mature humanity, remains under the spell of disaster unless it has a theory that can think the totality in its untruth. ... [T]his theory ... must incorporate societal and political reality and its dynamic" (CM 14/470). Under pressure from his account of late capitalism as an "exchange society," however, Adorno gives up a Hegelian understanding of society as an *articulated* totality whose internal differentiation has normative implications.

Previous chapters have emphasized four areas where Adorno's thought must be reassessed if the insights in his social philosophy are to be reclaimed: the basis of transformative hope, the scope of public authentication, the prospects for differential transformation, and the viability of new public spheres. Politically, all of these emphases share the need to rearticulate both the collective and the normative dimensions of Adorno's social philosophy. If this is not done, the disappointments of left-wing politics, together with transgressive individualism – a perennial fashion – will channel Adorno's contributions into a postmodern "politics" that, while pretending to be progressive, loses all connection with the Marxian core to Adorno's radical critique. It is appropriate, then, to revisit my claims in previous chapters with a view to their political implications.

6.2 SOCIAL ETHICS AND GLOBAL POLITICS

Implicit issues of collectivity and normativity are central to Chapter 2, where I try to recover Adorno's themes of suffering and hope but question how he interprets these themes. In particular, I question his turn to nonidentical objects as the basis for hoping that society as a whole can be transformed. This turn is, of course, a distinctive feature of Adorno's negative dialectic. If, as Hammer puts it, "negative dialectics *is* Adorno's political philosophy, though in a coded and highly mediated form,"[17] then criticisms of Adorno's turn to nonidentical objects require utmost care, lest one throw out the provocative baby with the problematic bathwater, to reuse a borrowed phrase from *Minima Moralia* (*MM* §22, p. 43/48). In a political context, however, such criticisms must be made. For the turn to nonidentical objects represents the absent collectivity and inarticulate normativity that trouble Adorno's political practice, political vision, and political theory. To highlight some political implications to my critical retrieval of Adorno's social philosophy, let me reflect further on recent debates concerning globalization.

Theory and Practice

The topic of globalization calls for scholarly approaches that are both normative and interdisciplinary. It is very difficult to achieve both at the same time. This has to do with the state of various academic disciplines. Social sciences that are more open to interdisciplinary perspectives tend to be normatively thin. Many of the humanities are ill-equipped to address comprehensive questions. And disciplines such as philosophy and theology that have more to say about normative issues often have little to offer on matters of societal structure and historical change. So we really do need the sort of comprehensive theorizing that Hegel provided with respect to modernity and that Horkheimer and Adorno attempted, albeit in a more fragmentary fashion, for the "dialectic of enlightenment."

Moreover, to take a critical stance toward debates about globalization requires a historically informed "ontology of the false condition" (*ND* 11/22) like the one Adorno modeled on Marx's critical theory of capitalism. A fundamental question framing these debates is whether globalization is occurring. The debate described by Held and McGrew between globalists (who say yes) and skeptics (who say no) illustrates the pitfalls of separating questions of historical process from questions of societal structure.[18] Globalists point to discrete factors that are new, without showing how together they make up a

---

[17] Hammer, *Adorno and the Political*, p. 98.
[18] See David Held and Anthony McGrew, *Globalization/Anti-Globalization* (Cambridge: Polity Press, 2002).

genuinely new historical phase. Skeptics point to large historical continuities without asking how trends at an earlier time might have a significantly different organization than "the same" trends today. To develop a satisfactory answer, and to provide an illuminating general description of globalization, one needs to provide both a historical and a structural account. Contemporary globalization needs to be understood as a distinct *historical phase* involving a significantly different *constellation of structural factors*. Hence, for example, the increasing dominance of the financial economy over production and distribution is a new structural feature of what some have labeled "turbocapitalism." Yet it is not a completely new feature. Rather it is a new outworking of a dynamic already identified by Marx in 1867 when he wrote: "The circulation of money as capital is ... an end in itself, for the expansion of value takes place only within this constantly renewed movement. The circulation of capital has therefore no limits."[19]

If not simply the economic predominance of finance capital but more generally the predominance of the economic dimension is a key to the contemporary shape of societal evil, and if, like Adorno, one rejects political resignation,[20] then one must seek a basis for hoping that both types of predominance can change. What would a sufficient basis include? Certainly, the existence of objects that resist the dictates of global capitalism – authentic artworks, perhaps, or not-yet-fully exploited creatures or unmet human needs – is not insignificant. Yet, as I argue in Chapter 2, these objects are not an adequate basis for transformative hope. We need, in addition, indications that a nonexploitative economy is historically possible, that viable spaces of organized resistance exist in contemporary society, and that collective interpretations of need provide substantial alternatives to turbocapitalist intensity. We need to reimagine legitimate

---

[19] Karl Marx, *Capital: A Critique of Political Economy*, vol. 1, ed. Frederick Engels, trans. Samuel Moore and Edward Aveling (New York: International Publishers, 1967), pp. 151–2.
[20] See Adorno's short and poignant "Resignation," in *CM* 289–93/794–9, one of the last essays he published before his untimely death in 1969. See also the essay that was originally intended to introduce the English translation of *Critical Models*: Henry Pickford, "The Dialectic of Theory and Practice: On Late Adorno," in Gibson and Rubin, *Adorno: A Critical Reader*, pp. 312–40.

roles for the state amid transnational frameworks of law and governance, lest transnational capital become in effect the final law of all lands. And we need ways to secure in a global civil society such visions of human flourishing as would foster the "transparent solidarity" that Adorno found lacking (*ND* 203–4/203–4).[21]

If multiple bases for transformative hope were articulated, we would not find ourselves in the desperate position where the praxis of theory is the only sufficiently good praxis to be theorized under contemporary conditions of societal evil. Here I have in mind passages such as *ND* 241–5/240–3, where Adorno proclaims the most advanced state of theory to be the only "authority [*Instanz*] for right practice and the good" (*ND* 242/240), and declares every individual or collective attempt to resist society as a whole to be "no less infected" by societal evil than someone "who does nothing at all" (*ND* 243/241). From this he draws a politically problematic conclusion: "[W]hoever cannot do anything without having it threaten to turn out bad, even if it aims for what is better, is compelled to think. This is the legitimation for thought and for intellectual satisfaction [*die des Glücks am Geiste*]." Paradoxically, then, a societal blockade on transformative praxis "gives thought a breathing spell that it would be a practical outrage not to use" (*ND* 245/243). In order to legitimate theory, Adorno seems compelled to privilege it as well. A better legitimation for theory, it seems to me, would be that many potential sources of change call for sufficiently comprehensive articulation. Not to attempt such an articulation would be to fail to contribute, in one of the ways that theory can, to the pursuit of human flourishing. There is no reason why theory itself must be the final arbiter of political practice. But there are many good reasons why theory should contribute to other forms of political practice, and why these in turn require theoretical articulation. Hence, Adorno's "strategic elitism" is not simply pragmatically inadequate, as Hammer seems to think. Because it misconstrues the proper relationship between theory and practice, it is politically illegitimate.

---

[21] On all of these topics, see Hauke Brunkhorst, *Solidarity: From Civic Friendship to a Global Legal Community*, trans. Jeffrey Flynn (Cambridge, Mass.: MIT Press, 2005). On "global civil society," see John Keane, *Global Civil Society?* (Cambridge: Cambridge University Press, 2003).

Social Democracy

That is one reason why Chapter 3 challenges Adorno's emphasis on "emphatic experience" and draws out its unexpected similarities with Heidegger's appeal to "authenticity." Although not antidemocratic in the way Heidegger is, Adorno has an untenable conception of democracy. For he thinks that the theorist's undemocratically established position in an insufficiently democratic society entitles or obligates the theorist to speak on behalf of those who are relatively unenlightened. I have already said why I find this conception problematic and why I think public authentication should be a hallmark of democratic truth telling. But let me comment briefly on how a more robust theory of democracy would eliminate the apparent need for Adorno's strategic elitism. For although Adorno assumes throughout his social philosophy that, in the modern context, a good society will be a democratic society, nowhere does he thematize either the general idea of democracy or the ranges of democratic institutions and practices that such a society would require.

Following John Dewey, I have claimed elsewhere that three concepts – freedom, participation, and recognition – are crucial to the general idea of democracy and that all three concepts are applicable to what can be distinguished as political, economic, and cultural democracy. Moreover, I have suggested that the prospects for political, economic, and cultural democracy are interlinked: in the long run, a society cannot have one without the others.[22] This provides a perspective from which to understand the paradoxes of contemporary globalization. While an emerging global civil society makes the prospects for cultural democracy ever more vivid, and while in rhetoric, although often not in practice, an increasing number of state and suprastate polities embrace political democracy, the turbocapitalist economy has become increasingly antidemocratic.

---

[22] Lambert Zuidervaart, "Postmodern Arts and the Birth of a Democratic Culture," in *The Arts, Community and Cultural Democracy*, ed. Lambert Zuidervaart and Henry Luttikhuizen (New York: St. Martin's, 2000), pp. 15–39. This essay's discussion of democracy draws primarily on John Dewey, *The Public and Its Problems*, in *The Later Works, 1925–1953*, vol. 2: *1925–1927*, ed. Jo Ann Boydston (Carbondale: Southern Illinois University Press, 1984), pp. 235–372. "Freedom, participation, and recognition" is my rearticulation of the French Revolution's *liberté, egalité, fraternité*.

In other words, current patterns of globalization exacerbate and extend a contradiction between democratic and antidemocratic tendencies that Hegel and Marx already identified in the structure of modern Western society. Whereas Adorno put his finger on the antidemocratic impetus of a late capitalist economy, he was unable to theorize the countervailing democratic forces both within this economy and within modern polities and cultures. Instead, he valorized authentic art and critical theory.

An instructive attempt to go beyond Adorno in this respect occurs in Hauke Brunkhorst's book *Solidarity*. Emphasizing reciprocations among cultural, economic, and political democracy, Brunkhorst identifies two "inclusion problems" under conditions of contemporary globalization: cultural fragmentation, which comes combined with fundamentalist conflicts, and social exclusion, which takes economic, sociocultural, and legal forms. These inclusion problems involve a failure to achieve democracy in all three respects. They "have put democracy on the defensive at the same moment that it appears to be triumphant worldwide and that, as a political idea and constitutional form, it stands virtually without alternative."[23] According to Brunkhorst, to solve both inclusion problems requires a democratization of global law, which currently tends to be hegemonic. That in turn relies both on the "globalization of civic solidarity" in an emerging transnational civil society, and on gains made toward securing human rights in international law. Brunkhorst also suggests that global democracy eventually will require something like a worldwide constitution and a transnational parliament. In this respect he agrees with cosmopolitan social democrats such as David Held and Anthony McGrew about the need for political democracy at the level of global governance. Yet Brunkhorst is more concerned than they are about the cultural sources of political democracy (via civil society) and its economic implications (via robust human rights).

My purpose here is neither to detail nor to debate Brunkhorst's argument, but rather to indicate how a full-fledged theory of social democracy would render Adorno's strategic elitism moot. If Brunkhorst is right about democratic potentials in transnational civil society and in

---

[23] Brunkhorst, *Solidarity*, p. 127. See also the discussion of Adorno's legacy in Hauke Brunkhorst, *Adorno and Critical Theory* (Cardiff: University of Wales Press, 1999).

internationally secured human rights, the contradiction between economic exclusion and cultural and political solidarity cannot stay hidden from public view. If this contradiction cannot stay hidden, then the critical theorist no longer needs to occupy a privileged position funded by publicly inaccessible emphatic experience. Emphatic experience may well remain crucial as a source of critical insight and political motivation, but it will no longer be what authenticates the truth of the critical theorist's claims.

Societal Disclosure

I do not wish to deny that eliminating social exclusion, as well as the suffering it causes, is extremely difficult to imagine. Contemporary patterns of social exclusion are not simply failures in democracy. They also involve illegitimate control and exploitation. Economic globalization further entrenches both the antidemocratic tendencies and the life-destroying effects of capitalism. Hence, efforts to curb or reverse these would require democratic arrangements that really do resist the ecological damage and economic exploitation both presupposed and fostered by contemporary globalization. To criticize these tendencies, and to imagine alternatives, one must appeal to substantive societal principles such as resourcefulness and justice and not simply to the procedural outworking of democratic norms having to do with freedom, participation, and recognition. Although the pursuit of justice under contemporary conditions requires a commitment to democracy, democracy per se neither fulfills the requirements of justice nor guarantees their fulfillment. Hence, the democratization of market economies is not an end in itself. Rather, the democratization of market economies is a necessary step toward ending economic exploitation and achieving an economy where resourcefulness prevails and neither justice nor solidarity is lacking: an economy where the earth receives proper care, where everyone's needs are met, and where no countries or groups or individuals have far more than they need while others suffer. That is why my retrieval of Adorno's critique of domination includes a normative critique of the capitalist economy.

The predominance of the economic dimension in contemporary globalization raises new questions about the relationship between functional differentiation and transfunctional integration. Some types of integration, while allowing for functional differentiation, can nevertheless hollow out differentiated spheres, destroy the earth, and foster oppression on a massive scale. This seems to characterize turbocapitalism. It provides integration that sparks and supports functional differentiation, but in a hollowing, destructive, and oppressive fashion. Social theorists such as Hauke Brunkhorst, who calls for "solidarity in the global legal community," and John Keane, who turns toward "global civil society," attempt to find alternative ways to integrate functionally differentiated societies without totally denying the contributions of turbocapitalism. Religionists such as Hans Küng and Rebecca Todd Peters take a different path, asking worldwide communities of faith, and anyone else who cares to listen, to help align society more fully with normative principles made available by traditions of faith. In other words, Brunkhorst and Keane seek a new structural integration; Küng and Peters look for a new normative integration. I think we need both – not only a new structural integration but also a new normative integration.

Accordingly, in Chapter 4 I call for a differential transformation of society. The idea of differential transformation supports a critique of contemporary globalization in three respects: with regard to different levels of interaction (social institutions, cultural practices, and interpersonal relations), in terms of structural interfaces (among economy, polity, and civil society), and with a view to societal principles such as resourcefulness, justice, and solidarity that are pervasive, distinct, and mutually complementary. Yet such a tripartite critique would be incomplete. The proposed evaluation with respect to different levels and structural interfaces emphasizes the need for proper *differentiation* but does not give an account of structural *integration*.[24] How are we to conceive of a proper structural integration under conditions of contemporary globalization?

Here the social vision proposed by Rebecca Todd Peters can help. She points to democratized power sharing, caring for the planet, and the social well-being of all people as a normative framework for

---

[24] The evaluation proposed in Chapter 4 with respect to societal principles is more complete, because it emphasizes both differentiation and normative integration.

evaluating stances toward globalization.[25] In my own terms, introduced in Chapter 3, a proper structural integration will have as its horizon a "life-giving disclosure" of society: a historical process "in which human beings and other creatures come to flourish, and not just some human beings or certain creatures, but all of them in their interconnections."[26] I agree with Peters that, in a contemporary setting, democratized power sharing is essential to the pursuit of such flourishing, and that earth keeping must go together with social well-being if the disclosure of society is to be truly life-giving. The idea of societal disclosure relativizes the achievements of functional differentiation. It encourages us to ask to what extent and in which respects the current array of differentiated levels and subsystems supports, promotes, hinders, or prevents the interconnected flourishing of human beings and other creatures. The idea also enables one to engage in an internal critique of specific differentiated levels or subsystems with a view to the larger historical process and structural constellation in which they participate and to which they contribute.

Clearly an articulation of societal principles in the direction of life-giving disclosure cannot simply appeal to emphatic experience. Nor can it be the prerogative of certain critical theorists. Rather it will emerge from the struggles of many groups and traditions to fashion and enact a "global ethic."[27] The notion of a global ethic implies in turn that an "ethics of resistance" (Espen Hammer) or an "ethical modernism" that appeals to "fugitive ethical events" (J. M. Bernstein) or a "politics of the mimetic shudder" (Martin Morris) would be insufficient as a critical retrieval of Adorno's social philosophy. For none of these appropriations address the questions of collective normativity that Adorno slighted and that a politics of globalization cannot avoid.

---

[25] Rebecca Todd Peters, *In Search of the Good Life: The Ethics of Globalization* (New York: Continuum, 2004), pp. 21–31.

[26] Lambert Zuidervaart, *Artistic Truth: Aesthetics, Discourse, and Imaginative Disclosure* (Cambridge: Cambridge University Press, 2004), p. 97.

[27] For an influential attempt by religious leaders from around the world, see Hans Küng and Karl-Josef Kuschel, eds., *A Global Ethic: The Declaration of the Parliament of the World's Religions* (New York: Continuum, 1998). See also the chapter "World Orders, Ethical Foundations," in Held and McGrew, *Globalization/Anti-Globalization*, pp. 88–97, and the chapter "Ethics beyond Borders" in Keane, *Global Civil Society?*, pp. 175–209.

Global Civil Society

The prospects for a global ethic hinge on the continuing emergence of a global civil society where economic alternatives and informal political publics can take root and thrive.[28] This is a larger political implication of the case made in Chapter 5 for the importance of new public spheres secured by a third-sector economy. For neither the arts nor any other fields of cultural endeavor can accomplish their distinctive tasks if they become soccer balls kicked back and forth between a corporate economy and administrative states or suprastates. Conversely, if an emerging global civil society does not secure transnationally the conditions of cultural endeavor, then there will be insufficient cross-cultural experience, conversation, and criticism to discover and articulate a global ethic.

I do not mean to argue, however, that the organizations and agencies of global civil society must be immunized from the corporate economy and state administration in order to have their own legitimacy. To advocate an immunization policy would be to fall into the "civil society purism" to which John Keane rightly objects. Tracing such purism back to Antonio Gramsci, Keane says that it views global civil society as "the *non-economic* space of social interaction 'located *between* the family, the state, and the market and operating *beyond* the confines of national societies, polities, and

---

[28] I prefer the term "informal" to "weak" political publics because I do not think that the activities and projects of cultural politics are necessarily less effective or transformative than those which have direct ties to governments and transnational governance. Rather, they have a different sort of political power, without which so-called strong publics (formal publics, in my own vocabulary) would themselves be ineffective. On the distinction between "weak" and "strong" publics see Nancy Fraser, *Justice Interruptus: Critical Reflections on the "Postsocialist" Condition* (New York: Routledge, 1997), pp. 69–98; Jürgen Habermas, *Between Facts and Norms: Contributions to a Discourse Theory of Law and Democracy*, trans. William Rehg (Cambridge, Mass.: MIT Press, 1996), pp. 302–14; Brunkhorst, *Solidarity*, pp. 137–42. In the "Translator's Introduction" to Brunkhorst's book, Jeffrey Flynn renders this distinction as one between weak and strong public spheres: "A public sphere is weak insofar as its deliberations shape opinion formation but have no power to make binding political decisions. ... A strong public sphere ... is authorized to make binding decisions – for example, parliaments and legislatures whose deliberations result in decisions enforced by state administrations" (p. xv).

economies.'"²⁹ Yet I think it is a category mistake, not to mention an error in political judgment, to cast the net so widely that global civil society includes *all* nongovernmental institutions and actors.³⁰ Keane makes this mistake when he defines global civil society as "*a dynamic nongovernmental system of interconnected socioeconomic institutions*" that includes, among other actors and agencies, "individuals, households, [and] profit-seeking businesses."³¹ By contrast, I would say certain nongovernmental actors and institutions, such as friendships and families, make up part of society in general but not civil society in particular, just as certain socioeconomic actors and institutions, especially for-profit businesses, although they are nongovernmental, do not belong to civil society as such. Moreover, individuals do not belong to one subsystem as distinct from another; rather, their activities occur within all subsystems.

To sort out such distinctions it is important to recognize the mixed character of the global economy. Just as, according to Chapter 5, the economy in North America has at least three sectors – a for-profit sector, a government economy, and a third sector of nonprofit, communal, and cooperative organizations – so the global economy is a mixed economy. This mixed character is important for both structural and normative reasons. Structurally, it provides a basis for maintaining interfaces among economy, polity, and civil society. Normatively, it provides ways to introduce a different vision of economy from that which prevails in the for-profit sector. One reason why ethical communities can make a difference in the process of globalization, as Küng and Peters clearly hope, is that many of their ventures are organized economically on a not-for-profit basis. This also suggests why the commercialization of such ventures is problematic: it undermines the economic not-for-profit basis upon which ethical communities could maintain their integrity and offer genuine alternatives.

For the most part, then, the primary economy of global civil society is neither the for-profit economy nor the governmental

---

[29] Keane, *Global Civil Society?*, p. 63, quoting from the editors' introduction to Helmut Anheier et al., *Global Civil Society 2001* (Oxford: Oxford University Press, 2001), p. 17.
[30] Here I concur with Jean Cohen and Andrew Arato, *Civil Society and Political Theory* (Cambridge, Mass.: MIT Press, 1992).
[31] Keane, *Global Civil Society?*, p. 8.

economy but a third-sector economy. Yet this does not mean that civil society has nothing to do with for-profit and government sectors of the economy. To argue for a wall of separation in the manner of "civil society purism" would be absurd: the most important international nongovernmental organizations (INGOs) rely heavily on revenues that come through government contracts, corporate donations, and grants from foundations whose assets stem from first-sector wealth (such as the Ford Foundation, the Rockefeller Foundation, and the Bill and Melinda Gates Foundation). Yet it would be a category mistake to include for-profit businesses in the fabric of civil society. It is more productive to think about the specific interfaces that allow money to flow from a for-profit economy to the third sector and that allow the goods and services generated in the third sector to recontextualize the operations of first-sector firms. In North America, corporate foundations, family foundations, and community foundations are pivotal in that regard. Similar attention should be paid to the specific interfaces that allow two-way traffic between a governmental economy and the third sector, such as tax breaks for charitable donations, and the third-sector provision of government-funded services in education and health care.

A nuanced understanding of global civil society and third-sector economics would also illuminate the institutional setting of social theory itself, a topic on which Critical Theory, whether Adornian or Habermasian, has contributed surprisingly little in recent years. Too many politically inflected social theories ignore how meshed universities have become with dominant economic and political systems and how vulnerable this makes higher education to corporate and state dictates. The switch since the 1970s from mostly state funding to increasing amounts of corporate funding is simply a shift *within* a paradigm that hinders genuine academic freedom. So we need to ask about the proper tasks of universities today, the extent to which they fulfill these tasks, the implications of their role for, say, ecology, health care, and cultural organizations, and the ways in which dedication to the university's proper tasks would contribute to life-giving disclosure. If economic and technological imperatives drive recent searches for new educational models, then this threatens the identity of universities as independent centers for research, critique, and cultural creation. Such an observation assumes, of course, that institutions of higher education do have proper tasks, that to fulfill these tasks they need to be independent from state or corporate control, and that they will fail in their tasks if they cave in to economic and technological imperatives.

Here my approach is quite different from John Keane's ambivalent celebration of the hybridized and fragmented multiversity that "cannot any longer pretend to speak through the grand narratives of Truth and Knowledge, but is instead confronted with the fact that it is a divided community that [contains both] rival conceptions of what constitutes success and competitor institutions ... that challenge its intellectual authority from outside."[32] How higher education along these lines can be "a principal catalyst and defender of global civil society and its ethos"[33] is a mystery to me. But I *can* see how independent universities dedicated to research, critique, and cultural creation could play a key role in the development of global civil society.

A different but related set of considerations pertains to distinctions between civil society proper and relationships of intimacy such as families, friendships, marriage, and what Rebecca Todd Peters calls "affectional communities." One way to understand the communitarian impulse is to say that it wishes both civil society and society in general to be more like relationships of intimacy. On that basis communitarians often decry the anonymity of the capitalist market and administrative states. When Peters connects civil society with "the rebuilding of community in the global North" and describes civil society as "a mobilization of grassroots aspects of the private sphere for the public good,"[34] she draws upon a communitarian understanding of civil society, even though she might not subscribe to a communitarian vision of politics and economics. I find such an approach problematic for two reasons: it misses the inherently public character of civil society organizations, and it expects more from affectional communities than they can properly deliver. In the effort to envision significant transformation, it overlooks the importance of functional differentiation. By contrast, although Keane's definition of civil society is too diffuse, his emphasis on "civility" as both a sectoral description and a guiding principle

---

[32] Ibid., p. 136.   [33] Ibid., p. 137.   [34] Peters, *In Search of the Good Life*, pp. 202–4.

has the advantage of emphasizing the public character of civil society organizations.

Hence, the politics implied by my critical retrieval of Adorno's social philosophy emphasizes both collectivity and normativity in a global context. It calls for an interdisciplinary, historically informed, and normative theory of globalization; an understanding of, and commitment to, political, economic, and cultural democracy; attention to processes of differentiation and integration that are both structural and normative; and the development of a global ethic within a global civil society secured by a not-for-profit economy. Although resistance to economic domination and exploitation remains crucial in such a politics, high culture and autonomous individuals are no longer the sole preserve of such resistance. Both the forces of resistance and the configuration of society turn out to be more collective in character and normative in orientation than an Adornian ethic would admit.

6.3 RESISTANCE AND TRANSFORMATION

Adorno scholars who regard ethics as his version of "first philosophy" (*prima philosophia*) might well object to my emphasis on collectivity and normativity in a democratic politics of global transformation.[35] Manuel Knoll, for example, argues that Adorno's "moral perspective" has primacy for his social theory and that this perspective turns on the themes of injustice and suffering as well as justice and happiness

---

[35] The description "first philosophy" derives from Manuel Knoll, *Theodor W. Adorno: Ethik als erste Philosophie* (Munich: Wilhelm Fink, 2002). Knoll's introduction (pp. 7–27) summarizes much of the German literature on Adorno's moral philosophy, including Mirko Wischke, *Kritik der Ethik des Gehorsams: Zum Moralproblem bei Theodor W. Adorno* (New York: Peter Lang, 1993); Gerhard Schweppenhäuser, *Ethik nach Auschwitz: Adornos negative Moralphilosophie* (Hamburg: Argument-Verlag, 1993); Gerhard Schweppenhäuser and Mirko Wischke, eds., *Impuls und Negativität: Ethik und Ästhetik bei Adorno* (Hamburg: Argument-Verlag, 1995); and Ulrich Kohlmann, *Dialektik der Moral: Untersuchungen zur Moralphilosophie Adornos* (Lüneburg: zu Klampen, 1997). In addition to Espen Hammer's *Adorno and the Political*, representative books in English on this topic include Drucilla Cornell, *The Philosophy of the Limit* (New York: Routledge, 1992); J. M. Bernstein, *Adorno: Disenchantment and Ethics* (Cambridge: Cambridge University Press, 2001); and Raymond Geuss, *Outside Ethics* (Princeton: Princeton University Press, 2005).

(*Glück*). In fact, Knoll says a "materialist and utopian hedonistic ethic" forms the core of Adorno's social thought.[36] An interpretation along these lines could fault my articulation of a global politics for ignoring Adorno's vigorous critique of existing forms of collectivity and his acute insights into contemporary dilemmas of normativity.

Adorno himself raises these issues when he identifies "the central problem of moral philosophy" as "the relationship of the ... particular human being and the universal that stands opposed to it." On the one hand, he says, contingent and "psychologically isolated" individuals can scarcely achieve "anything like freedom." On the other hand, "the abstract norm" is such that "living human beings" cannot "appropriate it for themselves in a living way." Both sides are "impossibilities," and the task of moral philosophy is to think through both in search of possible solutions (*PMP* 18–19/33–5). On the one hand, Adorno assumes that "the substantial nature of custom, the possibility of the good life in the forms in which the community exists ... has been radically eroded, that these forms have ceased to exist and that people today can no longer rely on them" (*PMP* 10/22). It is precisely for this reason that he prefers the term "morality" (*Moralität*) to "ethics" (*Sittlichkeit*) and follows Kant more than Hegel in his own moral philosophy. On the other hand, Adorno's thesis of the societal mediation of all individuals and their conduct makes him a harsh critic of an individualistic "ethics of authenticity," to borrow a phrase from Charles Taylor.[37] Because "moral and immoral conduct is always a social phenomenon," "it makes absolutely no sense to talk about ... moral conduct separately from relations of human beings to each other, and an individual who

---

[36] Knoll, *Adorno*, pp. 19–20. By insisting on the primacy of ethics for Adorno's social philosophy, political theory, epistemology, and aesthetics, Knoll takes issue with two other lines of interpretation: those who consider aesthetics to be Adorno's "first philosophy" (Wolfgang Welsch, Gerhard Kaiser, Jürgen Habermas, and Rüdiger Bubner), and those who interpret Adorno as primarily a religious or theological thinker (Mirko Wischke, Ulrich Kohlmann, and Helga Gripp). See pp. 22–24. Knoll does not discuss the substantial study by Hent de Vries, which was published in German in 1989 and has recently appeared in a revised English edition as *Minimal Theologies: Critiques of Secular Reason in Adorno and Levinas*, trans. Geoffrey Hale (Baltimore: Johns Hopkins University Press, 2005).

[37] Charles Taylor, *The Ethics of Authenticity* (Cambridge, Mass.: Harvard University Press, 1992). See Adorno, *PMP* 10–11/22–3, 13–15/26–8.

exists purely for himself is an empty abstraction" (*PMP* 19/34–5). Nevertheless, despite Adorno's critique of abstract individualism, an emphasis on collective normativity would seem problematic for his "negative moral philosophy"[38] or, better, for his metacritique of moral philosophy.[39] Whether one calls it a first philosophy or a last one, as Adorno preferred,[40] and whether one regards its primary topic to be ethics or morality, as Adorno also preferred, his reflections from damaged life seem incompatible with a transformative global politics.

If one's aim were simply to restate and defend Adorno's claims, such an objection would be irrefutable, despite the hermeneutical deficiency of any effort to reduce Adorno's social and political thought to a type of ethics. My aim is different, however, and on questions of collectivity and normativity I find Adorno's thought both theoretically inadequate and politically problematic. Yet my critical retrieval of Adorno's social philosophy does not ignore his insights. For as a critique of contemporary society that shows why both standard moral philosophies and customary ethical stances are inadequate "after Auschwitz," his social philosophy is unsurpassed. Even his reasons for questioning the slogan "What is to be done?" retain their relevance under contemporary conditions of globalization.

What I find missing in Adorno's thought – what in fact he deliberately resists – is anything like a social ethics. Nor does the ethical turn among many Adornian critical theorists address this lacuna. In fact their ethical turn misses something that *is* in Adorno's thought, namely, a dialectical affirmation of what I have called societal

---

[38] Schweppenhäuser, "Adorno's Negative Moral Philosophy."
[39] This description is suggested by the subtitle to the chapter or model on "Freedom" in *Negative Dialectics*: "On the Metacritique of Practical Reason" (*ND* 211/211). Adorno also uses the term "metacritique" in the title to his book on Husserlian phenomenology, cited in the next note, which has the misleading English translation *Against Epistemology*.
[40] "This is not a time for first philosophy, but for a last." Theodor W. Adorno, *Against Epistemology: A Metacritique; Studies in Husserl and the Phenomenological Antinomies* (1956), trans. Willis Domingo (Cambridge, Mass.: MIT Press, 1982), p. 40; *Zur Metakritik der Erkenntnistheorie: Studien über Husserl und die phänomenologischen Antinomien*, *Gesammelte Schriften* 5 (Frankfurt am Main: Suhrkamp, 1970), p. 47; translation modified. The sentence quoted concludes an introduction that criticizes every form of *prima philosophia*. This makes all the more puzzling attempts by later commentators to style Adorno's ethics as his own "first philosophy."

principles. Against those who would regard the universal as inherently oppressive, and would "attribute all the good to the individual," Adorno posits that "the universal always contains an implicit claim to represent a moral society in which force and compulsion have ceased to play any role" (*PMP* 18/34). This position seems to imply that principles such as justice and solidarity have a positive role to play both in society and in individual lives. Yet Adorno lacks an adequate account either of how the individual is constituted within multiple collectivities, and not simply within society as a whole, or of how such principles become effective in existing societies and do not merely represent what Derrida calls justice that "is yet to come."[41]

Gerhard Schweppenhäuser hints at this lack when he contrasts Adorno's moral philosophy with the "postmodern ethics" of Zygmunt Bauman. Schweppenhäuser claims that, unlike Adorno, Bauman fails to give an *"immanent* critique of modernity," problematically equates "philosophical universalism and imperialism," and thereby "negates moral-philosophical universalism abstractly."[42] By locating "the ultimate moral authority ... in the individual's moral intuition,"[43] Baumann ends up in an abstract individualism not unlike the ethics of authenticity that Adorno sharply criticized in the 1960s. Yet Adorno does not offer an adequate alternative, according to Schweppenhäuser: "The problem in Adorno's case is ... that he gives no theory of universalism but merely indications of its immanent dialectic."[44] Schweppenhäuser detects this problem in Adorno's refusal to provide a grounding or justification (*Begründung*) for the "new categorical imperative," which I have discussed in Chapter 2.

Perhaps returning to that discussion will help illuminate the lack not only in Adorno's moral philosophy but also in his politics. If we are *morally* obligated to arrange our thoughts and actions in such a way that nothing similar to Auschwitz will happen again (*ND* 365/358),

---

[41] "Justice remains, is yet, to come, *à venir*, it has an, it is *à-venir*, the very dimension of events irreducibly to come. It will always have it, this *à-venir*, and always has." Jacques Derrida, "Force of Law: The 'Mystical Foundation of Authority,'" in *Deconstruction and the Possibility of Justice*, ed. Drucilla Cornell, Michael Rosenfeld, and David Gray Carlson (New York: Routledge, 1992), pp. 3–67; quotation from p. 27.
[42] Schweppenhäuser, "Adorno's Negative Moral Philosophy," pp. 339, 341, 347.
[43] Ibid., p. 340.   [44] Ibid., p. 347.

then the question must arise whether this arrangement is possible and, if so, what it would be like. The question must arise because of a standard, albeit contested, assumption of moral philosophy since Kant that "ought implies can." Adorno shares this assumption when, for example, he poses the dilemma of how to judge the actions of those who perpetrated Nazi war crimes. A similar dilemma arises today in the zero-sum politics of terrorism and state-sponsored counterterrorism. Many of the Nazi perpetrators were either so psychologically damaged or externally coerced that they could not act upon their own moral judgments, if they had them. Yet their actions cried out for public justice. Moreover, the only way belatedly to bring about a measure of public justice itself seemed coercive. Adorno writes: "Here the latest stage of the moral dialectic comes to a head: acquittal would be a barefaced injustice, but a just reparation would be infected with the principle of brute force, and humanity [*Humanität*] is nothing but resistance to that" (*ND* 286/282). I would simply note that for this to be a "moral dialectic," the perpetrators must be considered morally and legally accountable. And to consider them accountable, we must assume with Adorno, and contrary to apparent fact, that they were able to act differently than they in fact did. In other words, "ought" implies "can."

But this assumption holds with respect to Adorno's new categorical imperative as well. We would only be morally obligated to prevent the repetition of Auschwitz if in fact we can. What would give us this ability? I find inadequate Adorno's appeal to a "materialistic motive" – the abhorrence of physical pain or, as he puts it prior to the previous quote, "the feeling of solidarity with what Brecht called 'tormentable bodies'" (*ND* 286/281). Adorno's appeal is inadequate for two reasons: first, as Chapter 2 suggests, because other culturally informed and ethically inflected feelings arise in the face of unspeakable suffering and, second, because the arrangements needed to prevent a repetition of Auschwitz go far beyond anything one individual or group or country can achieve. If ought implies can, then positing the new categorical imperative must assume that the necessary arrangements can in fact be achieved. This implies, in turn, that contemporary society is not the godforsaken desert Adorno paints in his bleaker moments. Otherwise the new categorical imperative would confront us with a massive moral gap not unlike the one my former colleague John Hare has identified in Kantian ethics.[45]

If we take seriously Adorno's new categorical imperative – if, indeed, we understand the moral dilemmas to which he points – then we should recover from Adornian neglect those moral sources and political agencies which would enable people to resist societal evil. Although not sufficient in themselves for individuals to "live rightly," such sources and agencies would at least keep people in touch with that within their lives and societies which is good. A social philosophy after Adorno requires the articulation of normative "universals" that are not abstract – societal principles such as justice, resourcefulness, and solidarity whose meaning neither floats in a modern heaven nor sinks into a postmodern morass but emerges historically "through clashes between societies and within them."[46] Also required is the realization that personal ethics and public morality are not enough, that neither resistant individuals nor steely imperatives will close multiple gaps between the historically emergent principles to which we aspire and the current arrangement of social institutions and cultural practices. To close the gaps will require concerted political efforts that, in transforming society, also reweave from within the fabric of our lives. To succeed, such a politics will need hues from the best colors available in the traditions that continue to sustain ethical communities around the world. The politics of transformation will need the rainbow of a truly global social ethic.

That is why, as theory, vision, and practice, Adorno's politics should not be reduced to a personal ethics of resistance. Rather, his ethics of resistance should be rearticulated within a democratic politics of global transformation. As Adorno once said, albeit with a different intent, "the quest for the good life is the quest for the right form of politics, if indeed such a right form of politics lay within the realm of what can be achieved today" (*PMP* 176/262). I am less skeptical than he about the possibility of an ethically attuned global politics. Such a politics would recognize collective agency, legitimate differentiation, and articulable norms. It would not abandon Adorno's emphasis on

---

[45] John E. Hare, *The Moral Gap: Kantian Ethics, Human Limits, and God's Assistance* (New York: Oxford University Press, 1996).
[46] Zuidervaart, *Artistic Truth*, p. 97.

suffering and hope. Neither would it ignore the societal evil identified by his critique of domination. Nor would it forget the need to fashion critical and utopian perspectives, perspectives that "displace and estrange the world, reveal it to be, with its rifts and crevices, as indigent and distorted as it will appear one day in the messianic light." To gain such perspectives without violence is both "the simplest of all things," says Adorno, and "utterly impossible" (*MM* §153, p. 247/283). I have suggested instead that doing so is far from simple, yet also not impossible. To insist that autonomous art and critical theory do not have the sole prerogative for fashioning the perspectives needed, and to show why gaining them without violence is not utterly impossible – these are the challenges facing a social philosophy after Adorno.

Appendix

Adorno's Social Philosophy

Theodor W. Adorno was one of the most important philosophers and social critics in Germany after World War II. Although less well known among Anglophone philosophers than his contemporary Hans-Georg Gadamer, Adorno had even greater influence on scholars and intellectuals in postwar Germany. In the 1960s he was the most prominent challenger to both Sir Karl Popper's philosophy of science and Martin Heidegger's philosophy of existence. Jürgen Habermas, Germany's foremost social philosopher after 1970, was Adorno's student and assistant. The scope of Adorno's influence stems from the interdisciplinary character of his research and of the Frankfurt School to which he belonged. It also stems from the thoroughness with which he examined Western philosophical traditions, especially from Kant onward, and the radicalness to his critique of contemporary Western society. He was a seminal social philosopher and a leading member of the first generation of Critical Theory.

Unreliable translations have hampered the reception of Adorno's published work in English-speaking countries. Since the 1990s, however, better translations have appeared, along with newly translated lectures and other posthumous works that are still being published. These materials not only facilitate an emerging assessment of his work in epistemology and ethics but also strengthen an already advanced reception of his work in aesthetics and cultural theory.

In this appendix I provide a brief sketch of Adorno's biography, followed by a summary of central topics and writings in his social philosophy.

Biographical Sketch

Born on September 11, 1903, as Theodor Ludwig Wiesengrund, Adorno lived in Frankfurt am Main for the first three decades of his life and the last two.[1] He was the only son of a wealthy German wine merchant of assimilated Jewish background and an accomplished musician of Corsican Catholic descent. Adorno studied philosophy with the neo-Kantian Hans Cornelius and music composition with Alban Berg. He completed his *Habilitationsschrift* on Kierkegaard's aesthetics in 1931, under the supervision of the Christian socialist Paul Tillich. After just two years as a university instructor (*Privatdozent*), he was expelled by the Nazis, along with other professors of Jewish heritage or on the political left. A few years later he turned his father's surname into a middle initial and adopted "Adorno," the maternal surname by which he is best known.

During the Nazi era Adorno resided in Oxford, New York City, and southern California. There he wrote several books for which he later became famous, including *Dialectic of Enlightenment* (with Max Horkheimer), *Philosophy of New Music*, *The Authoritarian Personality* (a collaborative project), and *Minima Moralia*. From these years come his provocative critiques of mass culture and the culture industry. Returning to Frankfurt in 1949 to take up a position in the philosophy department, Adorno quickly established himself as a leading German intellectual and a central figure in the Institute of Social Research. Founded as a freestanding center for Marxist scholarship in 1923, the institute had been led by Max Horkheimer since 1930. It provided the hub to what has come to be known as the Frankfurt School. Adorno became the institute's director in 1958. From the 1950s stem *In Search of Wagner*, Adorno's ideology-critique of the Nazi's

---

[1] This brief sketch derives from many sources. See in particular two comprehensive biographies that marked the centenary of Adorno's birth: Detlev Claussen, *Theodor W. Adorno: Ein letztes Genie* (Frankfurt am Main: Fischer, 2003); and Stefan Müller-Doohm, *Adorno: A Biography* (2003), trans. Rodney Livingstone (Cambridge: Polity Press, 2005).

favorite composer; *Prisms*, a collection of social and cultural studies; *Against Epistemology*, an antifoundationalist critique of Husserlian phenomenology; and the first volume of *Notes to Literature*, a collection of essays in literary criticism.

Conflict and consolidation marked the last decade of Adorno's life. A leading figure in the "positivism dispute" in German sociology, Adorno was a key player in debates about restructuring German universities and a lightning rod for both student activists and their right-wing critics. These controversies did not prevent him from publishing numerous volumes of music criticism, two more volumes of *Notes to Literature*, books on Hegel and on existential philosophy, and collected essays in sociology and in aesthetics. *Negative Dialectics*, Adorno's magnum opus on epistemology and metaphysics, appeared in 1966. *Aesthetic Theory*, the other magnum opus on which he had worked throughout the 1960s, appeared posthumously in 1970. He died of a heart attack on August 6, 1969, one month shy of his sixty-sixth birthday.

Dialectic of Enlightenment

Long before "postmodernism" became fashionable, Adorno and Horkheimer wrote one of the most searching critiques of modernity to have emerged among progressive European intellectuals. *Dialectic of Enlightenment* is a product of their wartime exile. It first appeared as a mimeograph titled *Philosophical Fragments* in 1944. This title became the subtitle when the book was published in 1947. Their book opens with a grim assessment of the modern West: "Enlightenment, understood in the widest sense as the advance of thought, has always aimed at liberating human beings from fear and installing them as masters. Yet the wholly enlightened earth radiates under the sign of disaster triumphant" (*DE* 1/25). How can this be, the authors ask. How can the progress of modern science and medicine and industry promise to liberate people from ignorance, disease, and brutal, mind-numbing work, yet help create a world where people willingly swallow fascist ideology, knowingly practice deliberate genocide, and energetically develop lethal weapons of mass destruction? Reason, they answer, has become irrational.

Although they cite Francis Bacon as a leading spokesman for an instrumentalized reason that becomes irrational, Horkheimer and Adorno do not think that modern science and scientism are the sole culprits. The tendency of rational progress to become irrational regress arises much earlier. For example, they cite both the Hebrew scriptures and Greek philosophers as contributing to regressive tendencies. If Horkheimer and Adorno are right, then a critique of modernity must also be a critique of premodernity, and a turn toward the postmodern cannot simply be a return to the premodern. Otherwise the failures of modernity will continue in a new guise under postmodern conditions. Society as a whole needs to be transformed.

Horkheimer and Adorno believe that society and culture form a historical totality, such that the pursuit of freedom in society is inseparable from the pursuit of enlightenment in culture (*DE* xvi/18). There is a flip side to this: a lack or loss of freedom in society – in the political, economic, and legal structures within which we live – signals a concomitant failure in cultural enlightenment – in philosophy, the arts, religion, and the like. The Nazi death camps are not an aberration, nor are mindless studio movies innocent entertainment. Both indicate that something fundamental has gone wrong in the modern West.

According to Horkheimer and Adorno, the source of today's disaster is a pattern of blind domination. As is discussed in Chapter 4, "domination" has a triple sense in this context: the domination of nature by human beings, the domination of nature within human beings, and, in both of these forms of domination, the domination of some human beings by others. What motivates such triple domination is an irrational fear of the unknown: "Humans believe themselves free of fear when there is no longer anything unknown. This has determined the path of demythologization. ... Enlightenment is mythical fear radicalized" (*DE* 11/38). In an unfree society whose culture pursues so-called progress no matter what the cost, that which is "other," whether human or nonhuman, gets shoved aside, exploited, or destroyed. The means of destruction may be more sophisticated in the modern West, and the exploitation may be less direct than outright slavery, but blind, fear-driven domination continues, with ever greater global consequences. The all-consuming engine driving this process is an ever-expanding capitalist economy, fed by scientific research and the latest technologies.

Contrary to some interpretations, Horkheimer and Adorno do not reject the eighteenth-century Enlightenment. Nor do they provide a negative "metanarrative" of universal historical decline. Rather, through a highly unusual combination of philosophical argument, sociological reflection, and literary and cultural commentary, they construct a "double perspective" on the modern West as a historical formation.[2] They summarize this double perspective in two interlinked theses: "Myth is already enlightenment, and enlightenment reverts to mythology" (*DE* xviii/21). The first thesis allows them to suggest that, despite being declared mythical and outmoded by the forces of secularization, older rituals, religions, and philosophies may have contributed to the process of enlightenment and may still have something worthwhile to contribute. The second thesis allows them to expose ideological and destructive tendencies within modern forces of secularization, but without denying either that these forces are progressive and enlightening or that the older conceptions they displace were themselves ideological and destructive.

A fundamental mistake in many interpretations of *Dialectic of Enlightenment* occurs when readers take such theses to be theoretical definitions of unchanging categories rather than critical judgments about historical tendencies. The authors are not saying that myth is "by nature" a force of enlightenment. Nor are they claiming that enlightenment "inevitably" reverts to mythology. In fact, what they find really mythical in both myth and enlightenment is the thought that fundamental change is impossible. Such resistance to change characterizes both ancient myths of fate and modern devotion to the facts.

Accordingly, in constructing a "dialectic of enlightenment" the authors simultaneously aim to carry out a dialectical enlightenment of enlightenment not unlike Hegel's *Phenomenology of Spirit*. Two Hegelian concepts anchor this project, namely, determinate negation and conceptual self-reflection. "Determinate negation" (*bestimmte Negation*) indicates that immanent criticism is the way to wrest truth

---

[2] Simon Jarvis, *Adorno: A Critical Introduction* (New York: Routledge, 1998), p. 23.

from ideology. A dialectical enlightenment of enlightenment "discloses each image as script. It teaches us to read from [the image's] features the admission of falseness which cancels its power and hands it over to truth" (*DE* 18/46). Beyond and through such determinate negation, a dialectical enlightenment of enlightenment also recalls the origin and goal of thought itself. Such recollection is the work of the concept as the self-reflection of thought (*der Begriff als Selbstbesinnung des Denkens*, *DE* 32/64). Conceptual self-reflection reveals that thought arises from the very corporeal needs and desires that get forgotten when thought becomes a mere instrument of human self-preservation. It also reveals that the goal of thought is not to continue the blind domination of nature and humans but to point toward reconciliation. Adorno works out the details of this conception in his subsequent lectures on Kant,[3] ethics,[4] and metaphysics[5] and in his books on Husserl,[6] Hegel,[7] and Heidegger.[8] His most comprehensive statement occurs in *Negative Dialectics*, which is discussed later.

Critical Social Theory

Dialectic of Enlightenment presupposes a critical social theory indebted to Karl Marx. Adorno reads Marx as a Hegelian materialist whose critique of capitalism unavoidably includes a critique of the ideologies that capitalism sustains and requires. The most important of these is what Marx called "the fetishism of commodities." Marx aims his critique of commodity fetishism against bourgeois social scientists who simply describe the capitalist economy but, in so doing, simultaneously misdescribe it and prescribe a false social vision. According to Marx, bourgeois economists necessarily ignore the exploitation

---

[3] *Kant's* Critique of Pure Reason (1959), ed. Rolf Tiedemann, trans. Rodney Livingstone (Stanford: Stanford University Press, 2001).
[4] *Problems of Moral Philosophy* (1963), ed. Thomas Schröder, trans. Rodney Livingstone (Stanford: Stanford University Press, 2000).
[5] *Metaphysics: Concept and Problems* (1965), ed. Rolf Tiedemann, trans. Edmund Jephcott (Stanford: Stanford University Press, 2000).
[6] *Against Epistemology: A Metacritique; Studies in Husserl and the Phenomenological Antinomies* (1956), trans. Willis Domingo (Cambridge, Mass.: MIT Press, 1982).
[7] *Hegel: Three Studies* (1963), trans. Shierry Weber Nicholsen (Cambridge, Mass.: MIT Press, 1993).
[8] *The Jargon of Authenticity* (1964), trans. Knut Tarnowski and Frederic Will (London: Routledge & Kegan Paul, 1973).

intrinsic to capitalist production. They fail to understand that capitalist production, for all its surface "freedom" and "fairness," must extract surplus value from the labor of the working class. Like ordinary producers and consumers under capitalist conditions, bourgeois economists treat the commodity as a fetish. They treat it as if it were a neutral object, with a life of its own, that directly relates to other commodities, in independence from the human interactions that actually sustain all commodities. Marx, by contrast, argues that whatever makes a product a commodity goes back to human needs, desires, and practices. The commodity would not have "use value" if it did not satisfy human wants. It would not have "exchange value" if no one wished to exchange it for something else. And its exchange value could not be calculated if the commodity did not share with other commodities a "value" created by the expenditure of human labor power and measured by the average labor time socially necessary to produce commodities of various sorts.

Adorno's social theory attempts to make Marx's central insights applicable to late capitalism. Although in agreement with Marx's analysis of the commodity, Adorno thinks his critique of commodity fetishism does not go far enough. Significant changes have occurred in the structure of capitalism since Marx's day. This requires revisions on a number of topics: the dialectic between forces of production and relations of production; the relationship between state and economy; the sociology of classes and class consciousness; the nature and function of ideology; and the role of expert cultures, such as modern art and social theory, in criticizing capitalism and calling for the transformation of society as a whole.

The primary clues to these revisions come from a theory of reification proposed by the Hungarian socialist Georg Lukács in the 1920s and from interdisciplinary projects and debates conducted by members of the Institute of Social Research in the 1930s and 1940s. Building on Max Weber's theory of rationalization, Lukács argues that the capitalist economy is no longer one sector of society alongside others. Rather, commodity exchange has become the central organizing principle for all sectors of society. This allows commodity fetishism to permeate all social institutions (e.g., law, administration, journalism) as well as all academic disciplines, including philosophy. "Reification" refers to "the structural process whereby the commodity form permeates life in capitalist society." Lukács was especially concerned with how reification makes human beings "seem like mere things obeying the inexorable laws of the marketplace."[9]

Initially Adorno shared this concern, even though he never had Lukács's confidence that the revolutionary working class could overcome reification. Later Adorno called the reification of consciousness an "epiphenomenon." What a critical social theory really needs to address is why hunger, poverty, and other forms of human suffering persist despite the technological and scientific potential to mitigate them or to eliminate them altogether. The root cause, Adorno says, lies in how capitalist relations of production have come to dominate society as a whole, leading to extreme, albeit often invisible, concentrations of wealth and power (*ND* 189–92/190–3). Society has come to be organized around the production of exchange values for the sake of producing exchange values, which, of course, always already requires a silent appropriation of surplus value. Adorno refers to this nexus of production and power as the "principle of exchange" (*Tauschprinzip*). A society where this nexus prevails is an "exchange society" (*Tauschgesellschaft*).

Adorno's diagnosis of the exchange society has three levels: politico-economic, social-psychological, and cultural. Politically and economically he responds to a theory of state capitalism proposed by Friedrich Pollock during the war years. An economist by training who was supposed to contribute a chapter to *Dialectic of Enlightenment* but never did,[10] Pollock argued that the state had acquired dominant economic power in Nazi Germany, the Soviet Union, and New Deal America. He called this new constellation of politics and economics "state capitalism." While acknowledging with Pollock that political and economic power have become more tightly meshed, Adorno does not think this fact changes the fundamentally economic character of capitalist exploitation. Rather, such exploitation has become even more abstract than it was in Marx's day, and therefore all the more effective and pervasive.

---

[9] Lambert Zuidervaart, *Adorno's Aesthetic Theory: The Redemption of Illusion* (Cambridge, Mass.: MIT Press, 1991), p. 76.
[10] Rolf Wiggershaus, *The Frankfurt School: Its History, Theories, and Political Significance*, trans. Michael Robertson (Cambridge, Mass.: MIT Press, 1994), pp. 313–19.

The social-psychological level in Adorno's diagnosis serves to demonstrate the effectiveness and pervasiveness of late capitalist exploitation. His American studies of anti-Semitism and the "authoritarian personality" argue that these pathologically extend "the logic of late capitalism itself, with its associated dialectic of enlightenment." People who embrace anti-Semitism and fascism tend to project their fear of abstract domination onto the supposed mediators of capitalism, while rejecting as elitist "all claims to a qualitative difference transcending exchange."[11]

Adorno's cultural studies show that a similar logic prevails in television, film, and the recording industries. In fact, Adorno first discovered late capitalism's structural change through his work with sociologist Paul Lazarsfeld on the Princeton (University) Radio Research Project. He articulated this discovery in a widely anthologized essay "On the Fetish-Character in Music and the Regression of Listening" (1938) and in "The Culture Industry," a chapter in *Dialectic of Enlightenment*. There, as I indicate in Chapter 5, Adorno argues that the culture industry dissolves the "genuine commodity character" that artworks possessed when exchange value still presupposed use value (*DE* 129–30/188). His main point is that culture-industrial hypercommercialization evidences a fateful shift in the structure of all commodities and therefore in the structure of capitalism itself.

Aesthetic Theory

Philosophical and sociological studies of the arts and literature make up more than half of Adorno's collected works (*Gesammelte Schriften*). All of his most important social-theoretical claims show up in these studies. Yet his "aesthetic writings" are not simply "applications" or "test cases" for theses developed in "nonaesthetic" texts. Adorno rejects any such separation of subject matter from methodology and all neat divisions of philosophy into specialized subdisciplines. This is one reason why academic specialists find his texts so challenging, not only musicologists and literary critics but also epistemologists and aestheticians.

[11] Jarvis, *Adorno*, p. 63.

First published the year after Adorno died, *Aesthetic Theory* marks the unfinished culmination of his remarkably rich body of aesthetic writings. It casts retrospective light on the entire corpus. It also comes closest to the model of "paratactical presentation" that Adorno, inspired especially by Walter Benjamin, found most appropriate for his own "atonal philosophy."[12] Relentlessly tracing concentric circles, *Aesthetic Theory* carries out a dialectical double reconstruction. It reconstructs the modern art movement from the perspective of philosophical aesthetics. It simultaneously reconstructs philosophical aesthetics, especially that of Kant and Hegel, from the perspective of modern art. From both sides, Adorno tries to elicit the sociohistorical significance of the art and philosophy discussed.

Adorno's claims about art in general stem from his reconstruction of the modern-art movement. So a summary of his philosophy of art sometimes needs to signal this by putting "modern" in parentheses. The book begins and ends with reflections on the social character of (modern) art. Two themes stand out in these reflections. One is an updated Hegelian question whether art can survive in a late capitalist world. The other is an updated Marxian question whether art can contribute to the transformation of this world. When addressing both questions, Adorno retains from Kant the notion that art proper ("fine art" or "beautiful art" – *schöne Kunst* – in Kant's vocabulary) is characterized by formal autonomy. But Adorno combines this Kantian emphasis on form with Hegel's emphasis on intellectual import (*geistiger Gehalt*) and Marx's emphasis on art's embeddedness in society as a whole. The result is a complex account of the simultaneous necessity and illusoriness of the artwork's autonomy. The artwork's necessary and illusory autonomy, in turn, is the key to (modern) art's social character, namely, to be "the social antithesis of society" (*AT* 8/19).

Adorno regards authentic works of (modern) art as social monads. The unavoidable tensions within them express unavoidable conflicts within the larger sociohistorical process from which they arise and to which they belong. These tensions enter the artwork through the

---

[12] See in this connection Robert Hullot-Kentor's "Translator's Introduction," *AT* xi–xxi.

artist's struggle with sociohistorically laden materials, and they call forth conflicting interpretations, many of which misread either the work-internal tensions or their connection to conflicts in society as a whole. Adorno sees all of these tensions and conflicts as "contradictions" to be worked through and eventually to be resolved. Their complete resolution, however, would require a transformation in society as a whole, which, given his social theory, does not seem imminent.

As commentary and criticism, Adorno's aesthetic writings are unparalleled in the subtlety and sophistication with which they trace work-internal tensions and relate them to unavoidable sociohistorical conflicts. One gets frequent glimpses of this in *Aesthetic Theory*. For the most part, however, the book proceeds at the level of "third reflections" – reflections on categories employed in actual commentary and criticism, with a view to their suitability for what artworks express and to their societal implications. Typically he elaborates these categories as polarities or dialectical pairs.

One such polarity, and a central one in Adorno's theory of artworks as social monads, occurs between the categories of import (*Gehalt*) and function (*Funktion*). Adorno's account of these categories distinguishes his sociology of art from both hermeneutical and empirical approaches. A hermeneutical approach would emphasize the artwork's inherent meaning or its cultural significance and downplay the artwork's political or economic functions. An empirical approach would investigate causal connections between the artwork and various social factors without asking hermeneutical questions about its meaning or significance. Adorno, by contrast, argues that, both as categories and as phenomena, import and function need to be understood in terms of each other. On the one hand, an artwork's import and its functions in society can be diametrically opposed. On the other hand, one cannot give a proper account of an artwork's social functions if one does not raise import-related questions about their significance. So too, an artwork's import embodies the work's social functions and has potential relevance for various social contexts. In general, however, and in line with his critiques of positivism and instrumentalized reason, Adorno gives priority to import, understood as societally mediated and socially significant meaning. The social functions emphasized in his own commentaries and criticisms are primarily intellectual functions rather than straightforwardly political or economic functions. This is consistent with the passage I use as an epigraph to Chapter 1, itself a hyperbolic version of the claim that (modern) art is society's social antithesis: "Insofar as a social function can be predicated for artworks, it is their functionlessness" (*AT* 227/336–7).

The priority of import also informs Adorno's stance on art and politics, which derives from debates with Lukács, Benjamin, and Bertolt Brecht in the 1930s. Because of the shift in capitalism's structure, and because of Adorno's own complex emphasis on (modern) art's autonomy, he doubts both the effectiveness and the legitimacy of tendentious, agitative, or deliberately consciousness-raising art. Yet he does see politically engaged art as a partial corrective to the bankrupt aestheticism of much mainstream art. Under the conditions of late capitalism, the best art, and politically the most effective, so thoroughly works out its own internal contradictions that the hidden contradictions in society can no longer be ignored. The plays of Samuel Beckett, to whom Adorno had intended to dedicate *Aesthetic Theory*, are emblematic in that regard. Adorno finds them more true than many other artworks.

Indeed, the idea of "truth content" (*Wahrheitsgehalt*) is the pivotal center around which all the concentric circles of Adorno's aesthetics turn.[13] To gain access to this center, one must temporarily suspend standard theories about the nature of truth (whether as correspondence, coherence, or pragmatic success) and allow for artistic truth to be dialectical, disclosive, and nonpropositional. According to Adorno, each artwork has its own import (*Gehalt*) by virtue of an internal dialectic between content (*Inhalt*) and form (*Form*). This import invites critical judgments about its truth or falsity. To do justice to the artwork and its import, such critical judgments need to grasp both the artwork's complex internal dynamics and the dynamics of the sociohistorical totality to which the artwork belongs. The artwork has an internal truth content to the extent that the

---

[13] See Albrecht Wellmer, "Truth, Semblance, Reconciliation: Adorno's Aesthetic Redemption of Modernity," in *The Persistence of Modernity: Essays on Aesthetics, Ethics, and Postmodernism*, trans. David Midgley (Cambridge, Mass.: MIT Press, 1991), pp. 1–35; Zuidervaart, *Adorno's Aesthetic Theory*; and Jarvis, *Adorno*, pp. 90–123.

artwork's import can be found internally and externally either true or false. Such truth content is not a metaphysical idea or essence hovering outside the artwork. But neither is it a merely human construct. It is historical but not arbitrary; nonpropositional, yet calling for propositional claims to be made about it; utopian in its reach, yet firmly tied to specific societal conditions. Truth content is the way in which an artwork simultaneously challenges the way things are and suggests how things could be better, but leaves things practically unchanged: "Art has truth as the semblance of the illusionless" (*AT* 132/199).

Negative Dialectics

Adorno's idea of artistic truth content presupposes the epistemological and metaphysical claims he works out most thoroughly in *Negative Dialectics*. These claims, in turn, consolidate and extend the historiographic and social-theoretical arguments already canvassed. As Simon Jarvis demonstrates, *Negative Dialectics* tries to formulate a "philosophical materialism" that is historical and critical but not dogmatic. Alternatively, one can describe the book as a "metacritique" of idealist philosophy, especially the philosophy of Kant and Hegel.[14] Adorno says the book aims "to use the strength of the [epistemic] subject to break through the deception [*Trug*] of constitutive subjectivity" (*ND* xx/10).

This occurs in four stages. First, a long introduction (*ND* 1–57/13–66) works out a concept of "philosophical experience" that both challenges Kant's distinction between "phenomena" and "noumena" and rejects Hegel's construction of "absolute spirit." Then part 1 (*ND* 59–131/67–136) distinguishes Adorno's project from the "fundamental ontology" in Heidegger's *Being and Time*. Part 2 (*ND* 133–207/137–207) works out Adorno's alternative with respect to the categories he reconfigures from German idealism. Part 3 (*ND* 209–408/209–400), composing nearly half the book, elaborates philosophical

---

[14] Jarvis, *Adorno*, pp. 148–74. See also the chapter titled "The Role of German Idealism in the Negative Dialectic," in Brian O'Connor, *Adorno's Negative Dialectic: Philosophy and the Possibility of Critical Rationality* (Cambridge, Mass.: MIT Press, 2004), pp. 15–43.

"models." These present negative dialectics in action upon key concepts of moral philosophy ("freedom"), philosophy of history ("world spirit" and "natural history"), and metaphysics. Adorno says the final model, devoted to metaphysical questions, "tries by critical self-reflection to give the Copernican revolution an axial turn" (*ND* xx/10). Alluding to Kant's self-proclaimed "second Copernican revolution," this description echoes Adorno's comment about breaking through the deception of constitutive subjectivity.

Like Hegel, Adorno criticizes Kant's distinction between phenomena and noumena by arguing that the transcendental conditions of experience can be neither so pure nor so separate from each other as Kant seems to claim. As concepts, for example, the a priori categories of the intellect or understanding (*Verstand*) would be unintelligible if they were not already about something that is nonconceptual. Conversely, the supposedly pure forms of space and time cannot simply be nonconceptual intuitions. Not even a transcendental philosopher would have access to them apart from concepts about them. So too, what makes possible any genuine experience cannot simply be the "application" of a priori concepts to a priori intuitions via the "schematism" of the imagination (*Einbildungskraft*). Genuine experience is made possible by that which exceeds the grasp of thought and sensibility. Adorno does not call this excess the "thing in itself," however, for that would assume the Kantian framework he criticizes. Rather, he calls it "the nonidentical" (*das Nichtidentische*).

The concept of the nonidentical, in turn, marks the difference between Adorno's materialism and Hegel's idealism. Although he shares Hegel's emphasis on a speculative identity between thought and being, between subject and object, and between reason and reality, Adorno denies that this identity has been achieved in a positive fashion. For the most part this identity has occurred negatively instead. That is to say, human thought, in achieving identity and unity, has imposed these upon objects, suppressing or ignoring their differences and diversity. Such imposition is driven by a societal formation whose exchange principle demands the equivalence (exchange value) of what is inherently nonequivalent (use value). Whereas Hegel's speculative identity amounts to an identity between identity and nonidentity, Adorno's amounts to a nonidentity between identity and nonidentity. That is why Adorno calls for a "negative dialectic" and why he rejects the affirmative character of Hegel's dialectic (*ND* 143–61/146–63).

Adorno does not reject the necessity of conceptual identification, however, nor does his philosophy claim to have direct access to the nonidentical. Under current societal conditions, thought can only have access to the nonidentical via conceptual criticisms of false identifications. Such criticisms must be "determinate negations," pointing up specific contradictions between what thought claims and what it actually delivers. Through determinate negation, those aspects of the object which thought misidentifies receive an indirect, conceptual articulation.

The motivation for Adorno's negative dialectic is not simply conceptual, however, nor are its intellectual resources. His epistemology is "materialist" in both regards. It is motivated, he says, by undeniable human suffering – a fact of unreason, if you will, to counter Kant's "fact of reason." Suffering is the corporeal imprint of society and the object upon human consciousness. The resources available to philosophy in this regard include the "expressive" or "mimetic" dimensions of language, which conflict with "ordinary" (i.e., societally sanctioned) syntax and semantics. In philosophy, this requires an emphasis on "presentation" (*Darstellung*) in which logical stringency and expressive flexibility interact (*ND* 18–19/ 29–31, 52–3/61–3). Another resource lies in unscripted relationships among established concepts. By taking such concepts out of their established patterns and rearranging them in "constellations" around a specific subject matter, philosophy can unlock some of the historical dynamic hidden within objects whose identity exceeds the classifications imposed upon them (*ND* 52–3/61–3, 162–6/164–8).

What unifies all of these desiderata, and what most clearly distinguishes Adorno's materialist epistemology from "idealism," whether Kantian or Hegelian, is his insisting on the "priority of the object" (*Vorrang des Objekts*, *ND* 183–97/184–97). Adorno regards as "idealist" any philosophy that affirms an identity between subject and object and thereby assigns constitutive priority to the epistemic subject. In insisting on the priority of the object, Adorno repeatedly makes three claims: first, that the epistemic subject is itself objectively constituted by the society to which it belongs and without which the subject could not exist; second, that no object can be fully known according to the rules and procedures of identitarian thinking; and, third, that the goal of thought itself, even when thought forgets its goal under societally induced pressures to impose identity on objects, is to honor them in their nonidentity, in their difference from what a restricted rationality declares them to be. Against empiricism, however, he argues that no object is simply "given" either, both because it can be an object only in relation to a subject and because objects are historical and have the potential to change.

Under current conditions the only way for philosophy to give priority to the object is dialectically, Adorno argues. He describes dialectics as the attempt to recognize the nonidentity between thought and the object while carrying out the project of conceptual identification. Dialectics is "the consistent consciousness of nonidentity," and contradiction, its central category, is "the nonidentical under the aspect of identity." Thought itself forces this emphasis on contradiction upon us, he says. To think is to identify, and thought can only achieve truth by identifying. So the semblance (*Schein*) of total identity lives within thought itself, mingled with thought's truth (*Wahrheit*). The only way to break through the semblance of total identity is immanently, using the concept. Accordingly, everything that is qualitatively different and that resists conceptualization will show up as a contradiction. "The contradiction is the nonidentical under the aspect of [conceptual] identity; the primacy of the principle of contradiction in dialectics tests the heterogeneous according to unitary thought [*Einheitsdenken*]. By colliding with its own boundary [*Grenze*], unitary thought surpasses itself. Dialectics is the consistent consciousness of nonidentity" (*ND* 5/17).

But thinking in contradictions is also forced upon philosophy by society itself. Society is riven with fundamental antagonisms, which, in accordance with the exchange principle, get covered up by identitarian thought. The only way to expose these antagonisms, and thereby to point toward their possible resolution, is to think against thought – in other words, to think in contradictions. In this way "contradiction" cannot be ascribed neatly to either thought or reality. Instead it is a "category of reflection" (*Reflexionskategorie*), enabling a thoughtful confrontation between concept (*Begriff*) and subject matter or object (*Sache*): "To proceed dialectically means to think in contradictions, for the sake of the contradiction already experienced in the object [*Sache*], and against that contradiction. A contradiction in reality, [dialectics] is a contradiction against reality" (*ND* 144–5/148).

The point of thinking in contradictions is not simply negative, however. It has a fragile, transformative horizon, namely, a society that would no longer be riven with fundamental antagonisms, thinking that would be rid of the compulsion to dominate through conceptual identification, and the flourishing of particular objects in their particularity. Because Adorno is convinced that contemporary society has the resources to alleviate the suffering it nevertheless perpetuates, his negative dialectics has a utopian reach: "In view of the concrete possibility of utopia, dialectics is the ontology of the false condition. A right condition would be freed from dialectics, no more system than contradiction" (*ND* 11/22). Such a "right condition" would be one of reconciliation between humans and nature, including the nature within human beings, and among human beings themselves. This idea of reconciliation sustains Adorno's reflections on ethics and metaphysics.

Ethics and Metaphysics

Like Adorno's epistemology, his moral philosophy derives from a materialistic metacritique of German idealism. The model on "Freedom" in *Negative Dialectics* (*ND* 211–99/211–94) conducts a metacritique of Kant's critique of practical reason. So too, the model on "World Spirit and Natural History" (*ND* 300–60/295–353) provides a metacritique of Hegel's philosophy of history. Both models simultaneously carry out a subterranean debate with the Marxist tradition, and this debate guides Adorno's appropriation of both Kantian and Hegelian "practical philosophy."

The first section in the introduction to *Negative Dialectics* indicates the direction Adorno's appropriation will take (*ND* 3–4/15–16). There he asks whether and how philosophy is still possible. Adorno asks this against the backdrop of Karl Marx's *Theses on Feuerbach*, which famously proclaimed that philosophy's task is not simply to interpret the world but to change it. In distinguishing his historical materialism from the sensory materialism of Ludwig Feuerbach, Marx portrays human beings as fundamentally productive and political organisms whose interrelations are not merely interpersonal but societal and historical. Marx's emphasis on production, politics, society, and history takes his epistemology in a "pragmatic" direction. "Truth" does not indicate the abstract correspondence between thought and reality, between proposition and fact, he says. Instead, "truth" refers to the economic, political, societal, and historical fruitfulness of thought in practice.

Although Adorno shares many of Marx's anthropological intuitions, he thinks that a twentieth-century equation of truth with practical fruitfulness had disastrous effects on both sides of the iron curtain. The introduction to *Negative Dialectics* begins by making two claims. First, although apparently obsolete, philosophy remains necessary because capitalism has not been overthrown. Second, Marx's interpretation of capitalist society was inadequate and his critique is outmoded. Hence, praxis no longer serves as an adequate basis for challenging (philosophical) theory. In fact, praxis serves mostly as a pretext for shutting down the theoretical critique that transformative praxis would require. Having missed the moment of its realization (via the proletarian revolution, according to early Marx), philosophy today must criticize itself: its societal naiveté, its intellectual antiquation, its inability to grasp the power at work in industrial late capitalism. While still pretending to grasp the whole, philosophy fails to recognize how thoroughly it depends upon society as a whole, all the way into philosophy's "immanent truth" (*ND* 4/16). Philosophy must shed such naiveté. It must ask, as Kant asked about metaphysics after Hume's critique of rationalism, How is philosophy still possible? More specifically, after the collapse of *Hegelian* thought, how is philosophy still possible? How can the dialectical effort to conceptualize the nonconceptual – which Marx also pursued – how can this philosophy be continued?

This self-implicating critique of the relation between theory and practice is one crucial source to Adorno's reflections on ethics and metaphysics. Another is the catastrophic impact of twentieth-century history on the prospects for imagining and achieving a more humane world. Adorno's is an ethics and metaphysics "after Auschwitz." Ethically, he says, Hitler's barbarism imposes a "new categorical imperative" on human beings in their condition of unfreedom: so to arrange their thought and action that "Auschwitz would not repeat itself, [that] nothing similar would happen" (*ND* 365/358). Metaphysically, philosophers must find historically appropriate ways to speak about meaning and truth and suffering that neither deny nor affirm the existence of a world transcendent to the one we know. Whereas denying it would suppress the suffering that calls out for fundamental change, straightforwardly affirming the existence of utopia would cut off the critique of contemporary society and the struggle to change it. The basis for Adorno's double strategy is not a hidden ontology, as some have suggested, but rather a "speculative" or "metaphysical" experience. Adorno appeals to the experience of unconstrained thought's envisioning a world where current suffering would be abolished and past suffering revoked (*ND* 403/395). Neither logical positivist antimetaphysics nor Heideggerian hypermetaphysics can do justice to this experience.

Adorno indicates his own alternative to both traditional metaphysics and more recent antimetaphysics in passages that juxtapose resolute self-criticism and impassioned hope. His historiographic, social theoretical, aesthetic, and negative dialectical concerns meet in passages such as the one quoted at greater length in Chapter 2: "Thought that does not capitulate before wretched existence comes to naught before its criteria, truth becomes untruth, philosophy becomes folly. And yet philosophy cannot give up, lest idiocy triumph in actualized unreason [*Widervernunft*]. ... Folly is truth in the shape that human beings must accept whenever, amid the untrue, they do not give up truth" (*ND* 404/396). If the ongoing assessment of Adorno's social philosophy does not address such passages, it will not truly have begun.

Bibliography

Theodor W. Adorno

This section is divided into two parts: the first lists many of Adorno's books in English, as well as those volumes in German which are cited in this book; the second lists some anthologies of Adorno's writings in English and essays cited in this book. A date in parentheses following a title indicates either the first German edition or, in the case of posthumous publications, the date of the original lectures. The abbreviation *GS* or *NS* after an entry tells where a book can be found in Adorno's collected writings. *GS* indicates writings published during Adorno's lifetime and collected in the twenty volumes of Theodor W. Adorno, *Gesammelte Schriften*, edited by Rolf Tiedemann et al. (Frankfurt am Main: Suhrkamp, 1970–86). *NS* indicates posthumous works that are appearing as editions of the Theodor W. Adorno Archive in the collection *Nachgelassene Schriften* (Frankfurt am Main: Suhrkamp, 1993–). For more extensive Adorno bibliographies, see the biography by Stefan Müller-Doohm (2005) and the anthology edited by Tom Huhn (2004), both cited in the subsequent section listing works by other authors.

Books

Aesthetic Theory (1970). Translated, edited, and with a translator's introduction by Robert Hullot-Kentor. Minneapolis: University of Minnesota Press, 1997. (*GS* 7)

Ästhetische Theorie. Gesammelte Schriften 7. Edited by Gretel Adorno and Rolf Tiedemann. 2d ed. Frankfurt am Main: Suhrkamp, 1972.
Against Epistemology: A Metacritique; Studies in Husserl and the Phenomenological Antinomies (1956). Translated by Willis Domingo. Cambridge, Mass.: MIT Press, 1982. (*GS* 5)
The Authoritarian Personality. By T.W. Adorno et al. New York: Harper & Brothers, 1950. (*GS* 9.1)
Critical Models: Interventions and Catchwords (1963, 1969). Translated by Henry W. Pickford. New York: Columbia University Press, 1998. (*GS* 10.2)
Dialectic of Enlightenment: Philosophical Fragments (1947). By Max Horkheimer and Theodor W. Adorno. Edited by Gunzelin Schmid Noerr. Translated by Edmund Jephcott. Stanford: Stanford University Press, 2002. (*GS* 3)
Dialektik der Aufklärung. In Max Horkheimer, *Gesammelte Schriften, Band 5: "Dialektik der Aufklärung" und Schriften 1940–1950.* Edited by Gunzelin Schmid Noerr. Frankfurt am Main: Fischer Taschenbuch, 1987.
Hegel: Three Studies (1963). Translated by Shierry Weber Nicholsen. Cambridge, Mass.: MIT Press, 1993. (*GS* 5)
In Search of Wagner (1952). Translated by Rodney Livingstone. London: NLB, 1981. (*GS* 13)
Introduction to Sociology (1968). Edited by Christoph Gödde. Translated by Edmund Jephcott. Stanford: Stanford University Press, 2000. (*NS* IV.15)
The Jargon of Authenticity (1964). Translated by Knut Tarnowski and Frederic Will. London: Routledge & Kegan Paul, 1973. (*GS* 6)
Jargon der Eigentlichkeit: Zur deutschen Ideologie. Gesammelte Schriften 6. Frankfurt am Main: Suhrkamp, 1973.
Kant's Critique of Pure Reason (1959). Edited by Rolf Tiedemann. Translated by Rodney Livingstone. Stanford: Stanford University Press, 2001. (*NS* IV.4)
Kierkegaard: Construction of the Aesthetic (1933). Translated by Robert Hullot-Kentor. Minneapolis: University of Minnesota Press, 1989. (*GS* 2)
Kierkegaard: Konstruktion des Ästhetischen. Gesammelte Schriften 2. Frankfurt am Main: Suhrkamp, 1979.
Metaphysics: Concept and Problems (1965). Edited by Rolf Tiedemann. Translated by Edmund Jephcott. Stanford: Stanford University Press, 2000. (*NS* IV.14)
Minima Moralia: Reflections from Damaged Life (1951). Translated by E.F.N. Jephcott. London: NLB, 1974. (*GS* 4)
Minima Moralia: Reflexionen aus dem beschädigten Leben. Gesammelte Schriften 4. 2d ed. Frankfurt am Main: Suhrkamp, 1996.
Negative Dialectics (1966). Translated by E.B. Ashton. New York: Seabury Press, 1973. (*GS* 6)
Negative Dialektik. Gesammelte Schriften 6. Frankfurt am Main: Suhrkamp, 1973.

Notes to Literature (1958, 1961, 1965, 1974). 2 vols. Edited by Rolf Tiedemann. Translated by Shierry Weber Nicholsen. New York: Columbia University Press, 1991, 1992. (*GS* 11)
Philosophy of New Music (1949). Translated, edited, and with an introduction by Robert Hullot-Kentor. Minneapolis: University of Minnesota Press, 2006. (*GS* 12)
The Positivist Dispute in German Sociology (1969). By Theodor W. Adorno et al. Translated by Glyn Adey and David Frisby. London: Heinemann, 1976. (*GS* 8)
Prisms (1955). Translated by Samuel Weber and Shierry Weber. London: Neville Spearman, 1967; Cambridge, Mass.: MIT Press, 1981. (*GS* 10.1)
Probleme der Moralphilosophie (1963). Edited by Thomas Schröder. Frankfurt am Main: Suhrkamp, 1996.
Problems of Moral Philosophy (1963). Edited by Thomas Schröder. Translated by Rodney Livingstone. Stanford: Stanford University Press, 2000. (*NS* IV.10)
Soziologische Schriften I. Gesammelte Schriften 8. Frankfurt am Main: Suhrkamp, 1972.
Zur Metakritik der Erkenntnistheorie: Studien über Husserl und die phänomenologischen Antinomien. Gesammelte Schriften 5. Frankfurt am Main: Suhrkamp, 1970.

Anthologies and Essays

The Adorno Reader. Edited by Brian O'Connor. Oxford: Blackwell, 2000.
"Alienated Masterpiece: The *Missa Solemnis* (1959)." *Telos*, no. 28 (1976): 113–24.
Can One Live after Auschwitz? A Philosophical Reader. Edited by Rolf Tiedemann. Translated by Rodney Livingstone et al. Stanford: Stanford University Press, 2003.
The Culture Industry: Selected Essays on Mass Culture. Edited by J. M. Bernstein. London: Routledge, 1991.
"Erinnerungen an Paul Tillich." In *Werk und Wirken Paul Tillichs: Ein Gedenkbuch*, pp. 24–38. Stuttgart: Evangelisches Verlagswerk, 1967.
Essays on Music: Theodor W. Adorno. Edited by Richard D. Leppert. Translated by Susan H. Gillespie et al. Berkeley: University of California Press, 2002.

Other Authors

Aho, Kevin. "Why Heidegger Is Not an Existentialist: Interpreting Authenticity and Historicity in *Being and Time*." *Florida Philosophical Review* 3, no. 2 (Winter 2003): 5–22.
Anker, Roy, ed. *Dancing in the Dark: Youth, Popular Culture, and the Electronic Media*. Grand Rapids, Mich.: Eerdmans, 1991.

Baumeister, Thomas, and Jens Kulenkampff. "Geschichtsphilosophie und philosophische Ästhetik. Zu Adornos 'Ästhetischer Theorie.'" *Neue Hefte für Philosophie*, no. 5 (1973): 74–104.
Becker-Schmidt, Regina. "Critical Theory as a Critique of Society: Theodor W. Adorno's Significance for a Feminist Sociology." In *Adorno, Culture and Feminism*, edited by Maggie O'Neill, pp. 104–18. London: Sage, 1999.
Benhabib, Seyla. *Critique, Norm, and Utopia: A Study of the Foundations of Critical Theory*. New York: Columbia University Press, 1986.
Benjamin, Jessica. "The End of Internalization: Adorno's Social Psychology." *Telos*, no. 32 (Summer 1977): 42–64.
Benjamin, Walter. "Das Kunstwerk im Zeitalter seiner technischen Reproduzierbarkeit." In *Illuminationen: Ausgewählte Schriften*, pp. 136–69. Frankfurt am Main: Suhrkamp, 1977.
   "The Work of Art in the Age of Mechanical Reproduction." In *Illuminations*, edited by Hannah Arendt, translated by Harry Zohn, pp. 217–51. New York: Schocken Books, 1969.
Berman, Russell. "Adorno's Politics." In *Adorno: A Critical Reader*, edited by Nigel Gibson and Andrew Rubin, pp. 110–31. Oxford: Blackwell, 2002.
Bernstein, J. M. *Adorno: Disenchantment and Ethics*. Cambridge: Cambridge University Press, 2001.
   "Negative Dialectic as Fate: Adorno and Hegel." In *The Cambridge Companion to Adorno*, edited by Tom Huhn, pp. 19–50. Cambridge: Cambridge University Press, 2004.
Bozzetti, Mauro. "Hegel on Trial." In *Adorno: A Critical Reader*, edited by Nigel Gibson and Andrew Rubin, pp. 292–311. Oxford: Blackwell, 2002.
Brand, Peggy Zeglin. "Revising the Aesthetic-Nonaesthetic Distinction: The Aesthetic Value of Activist Art." In *Feminism and Tradition in Aesthetics*, edited by Peggy Zeglin Brand and Carolyn Korsmeyer, pp. 245–72. University Park: Pennsylvania State University Press, 1995.
Brantlinger, Patrick. *Bread and Circuses: Theories of Mass Culture as Social Decay*. Ithaca: Cornell University Press, 1983.
Brunkhorst, Hauke. *Adorno and Critical Theory*. Cardiff: University of Wales Press, 1999.
   Solidarity: From Civic Friendship to a Global Legal Community. Translated by Jeffrey Flynn. Cambridge, Mass.: MIT Press, 2005.
Bürger, Peter. "Critique of Autonomy." In *Encyclopedia of Aesthetics*, edited by Michael Kelly, 1:175–8. New York: Oxford University Press, 1998.
   Theory of the Avant-Garde. Translated by Michael Shaw, with a foreword by Jochen Schulte-Sass. Minneapolis: University of Minnesota Press, 1984.
Carman, Taylor. *Heidegger's Analytic: Interpretation, Discourse, and Authenticity in Being and Time*. Cambridge: Cambridge University Press, 2003.
Carroll, Noël. "Essence, Expression, and History: Arthur Danto's Philosophy of Art." In *Danto and His Critics*, edited by Mark Collins, pp. 79–106. Oxford: Blackwell, 1993.

A Philosophy of Mass Art. Oxford: Clarendon, 1998.
Claussen, Detlev. *Theodor W. Adorno: Ein letztes Genie*. Frankfurt am Main: Fischer, 2003.
Cohen, Jean, and Andrew Arato. *Civil Society and Political Theory*. Cambridge, Mass.: MIT Press, 1992.
Cook, Deborah. *Adorno, Habermas, and the Search for a Rational Society*. New York: Routledge, 2004.
  "Adorno on Mass Societies." *Journal of Social Philosophy* 32 (Spring 2001): 35–52.
  The Culture Industry Revisited: Theodor W. Adorno on Mass Culture. Lanham, Md.: Rowman & Littlefield, 1996.
  "From the Actual to the Possible: Nonidentity Thinking." *Constellations* 12 (2005): 21–35.
Cornell, Drucilla. *The Philosophy of the Limit*. New York: Routledge, 1992.
Dahlstrom, Daniel O. *Heidegger's Concept of Truth*. Cambridge: Cambridge University Press, 2001.
Danto, Arthur C. *After the End of Art: Contemporary Art and the Pale of History*. Princeton: Princeton University Press, 1997.
  The Philosophical Disenfranchisement of Art. New York: Columbia University Press, 1986.
  The Transfiguration of the Commonplace: A Philosophy of Art. Cambridge, Mass.: Harvard University Press, 1981.
  The Wake of Art: Criticism, Philosophy, and the Ends of Taste. Edited by Greg Horowitz and Tom Huhn. Amsterdam: G + B Arts International, 1998.
Derrida, Jacques. "Force of Law: The 'Mystical Foundation of Authority.'" In *Deconstruction and the Possibility of Justice*, edited by Drucilla Cornell, Michael Rosenfeld, and David Gray Carlson, pp. 3–67. New York: Routledge, 1992.
Devereaux, Mary. "Autonomy and Its Feminist Critics." In *Encyclopedia of Aesthetics*, edited by Michael Kelly, 1:178–82. New York: Oxford University Press, 1998.
  "The Philosophical and Political Implications of the Feminist Critique of Aesthetic Autonomy." In *Turning the Century: Feminist Criticism in the 1990s*, edited by Glynis Carr, pp. 164–86. Lewisburg, Pa.: Bucknell University Press, 1992.
  "Protected Space: Politics, Censorship, and the Arts." *Journal of Aesthetics and Art Criticism* 51 (Spring 1993): 207–15.
Dewey, John. *The Public and Its Problems*. In *The Later Works, 1925–1953*, Vol. 2: *1925–1927*, edited by Jo Ann Boydston, pp. 235–372. Carbondale: Southern Illinois University Press, 1984.
Dreyfus, Hubert L. *Being-in-the-World: A Commentary on Heidegger's* Being and Time, *Division I*. Cambridge, Mass.: MIT Press, 1991.
Duvenage, Pieter. *Habermas and Aesthetics: The Limits of Communicative Reason*. Malden, Mass.: Polity Press, 2003.
Felski, Rita. *Beyond Feminist Aesthetics: Feminist Literature and Social Change*. Cambridge, Mass.: Harvard University Press, 1989.

"Why Feminism Doesn't Need an Aesthetic (And Why It Can't Ignore Aesthetics)." In *Feminism and Tradition in Aesthetics*, edited by Peggy Zeglin Brand and Carolyn Korsmeyer, pp. 431–45. University Park: Pennsylvania State University Press, 1995.

Fraser, Nancy. *Justice Interruptus: Critical Reflections on the "Postsocialist" Condition*. New York: Routledge, 1997.

Geuss, Raymond. *Outside Ethics*. Princeton: Princeton University Press, 2005.

Gibson, Nigel, and Andrew Rubin, eds. *Adorno: A Critical Reader*. Oxford: Blackwell, 2002.

Goudzwaard, Bob. *Capitalism and Progress: A Diagnosis of Western Society*. Translated and edited by Josina Van Nuis Zylstra. Grand Rapids, Mich.: Eerdmans, 1979.

Gracyk, Theodore. *Rhythm and Noise: An Aesthetics of Rock*. London: I. B. Tauris, 1996.

Grenz, Friedemann. *Adornos Philosophie in Grundbegriffen. Auflösung einiger Deutungsprobleme*. Frankfurt am Main: Suhrkamp, 1974.

Guignon, Charles. "Philosophy and Authenticity: Heidegger's Search for a Ground for Philosophizing." In *Heidegger, Authenticity, and Modernity: Essays in Honor of Hubert L. Dreyfus, Volume 1*, edited by Mark A. Wrathall and Malpas Jeff, pp. 79–101. Cambridge, Mass.: MIT Press, 2000.

Habermas, Jürgen. *Autonomy and Solidarity: Interviews with Jürgen Habermas*. Edited by Peter Dews. Rev. ed. London: Verso, 1992.

——— *Between Facts and Norms: Contributions to a Discourse Theory of Law and Democracy*. Translated by William Rehg. Cambridge, Mass.: MIT Press, 1996.

——— "Concluding Remarks." In *Habermas and the Public Sphere*, edited by Craig Calhoun, pp. 462–79. Cambridge, Mass.: MIT Press, 1992.

——— "Further Reflections on the Public Sphere." In *Habermas and the Public Sphere*, edited by Craig Calhoun, pp. 421–61. Cambridge, Mass.: MIT Press, 1992.

——— *Knowledge and Human Interests*. Translated by Jeremy J. Shapiro. Boston: Beacon Press, 1971.

——— *Moral Consciousness and Communicative Action*. Translated by Christian Lenhardt and Shierry Weber Nicholsen. Introduction by Thomas McCarthy. Cambridge, Mass.: MIT Press, 1990.

——— *The Philosophical Discourse of Modernity: Twelve Lectures*. Translated by Frederick G. Lawrence. Cambridge, Mass.: MIT Press, 1987.

——— *Der philosophische Diskurs der Moderne: Zwölf Vorlesungen*. Frankfurt am Main: Suhrkamp, 1985.

——— *Philosophical-Political Profiles*. Translated by Frederick G. Lawrence. Cambridge, Mass.: MIT Press, 1983.

——— *Philosophisch-politische Profile*. Frankfurt am Main: Suhrkamp, 1971.

——— *Postmetaphysical Thinking: Philosophical Essays*. Translated by William Mark Hohengarten. Cambridge, Mass.: MIT Press, 1992.

"The Public Sphere: An Encyclopedia Article" (1964). Translated by Sara Lennox and Frank Lennox. *New German Critique*, no. 1 (Fall 1974): 49–55.
The Structural Transformation of the Public Sphere: An Inquiry into a Category of Bourgeois Society. Translated by Thomas Burger and Frederick Lawrence. Cambridge, Mass.: MIT Press, 1989.
Theorie des kommunikativen Handelns. 2 vols. Frankfurt am Main: Suhrkamp, 1981.
The Theory of Communicative Action. Translated by Thomas McCarthy. 2 vols. Boston: Beacon Press, 1984, 1987.
Toward a Rational Society: Student Protest, Science, and Politics. Translated by Jeremy J. Shapiro. Boston: Beacon Press, 1970.
Truth and Justification. Edited and with translations by Barbara Fultner. Cambridge, Mass.: MIT Press, 2003.
Hammer, Espen. "Adorno and Extreme Evil." *Philosophy and Social Criticism* 26, no. 4 (2000): 75–93.
Adorno and the Political. New York: Routledge, 2005.
Hansen, Miriam. "Mass Culture as Hieroglyphic Writing: Adorno, Derrida, Kracauer." *New German Critique*, no. 56 (Spring–Summer 1992): 43–73.
Hare, John E. *The Moral Gap: Kantian Ethics, Human Limits, and God's Assistance.* New York: Oxford University Press, 1996.
Harvey, David. *The New Imperialism.* Oxford: Oxford University Press, 2003.
Haskins, Casey. "Autonomy: Historical Overview." In *Encyclopedia of Aesthetics*, edited by Michael Kelly, 1:170–5. New York: Oxford University Press, 1998.
Heberle, Renée J., ed. *Feminist Interpretations of Theodor Adorno.* University Park: Pennsylvania State University Press, 2006.
Hegel, G. W. F. *Elements of the Philosophy of Right.* Edited by Allen W. Wood. Translated by H. B. Nisbet. Cambridge: Cambridge University Press, 1991.
Phenomenology of Spirit. Translated by A. V. Miller. Oxford: Oxford University Press, 1977.
Heidegger, Martin. *Being and Time.* Translated by John Macquarrie and Edward Robinson. New York: Harper & Row, 1962.
Being and Time. Translated by Joan Stambaugh. Albany: State University of New York Press, 1996.
Sein und Zeit. 15th ed. Tübingen: Max Niemeyer, 1979.
Held, David, and Anthony McGrew. *Globalization/Anti-Globalization.* Cambridge: Polity Press, 2002.
Honneth, Axel. "Communication and Reconciliation: Habermas' Critique of Adorno." *Telos*, no. 39 (Spring 1979): 45–61.
The Critique of Power: Reflective Stages in a Critical Social Theory. Translated by Kenneth Baynes. Cambridge, Mass.: MIT Press, 1991.
The Fragmented World of the Social: Essays in Social and Political Philosophy. Edited by Charles W. Wright. Albany: State University of New York Press, 1995.

"The Possibility of a Disclosing Critique of Society: The *Dialectic of Enlightenment* in Light of Current Debates in Social Criticism." *Constellations* 7, no. 1 (2000): 116–27.

Huhn, Tom, ed. *The Cambridge Companion to Adorno*. Cambridge: Cambridge University Press, 2004.

Huhn, Tom, and Lambert Zuidervaart, eds. *The Semblance of Subjectivity: Essays in Adorno's Aesthetic Theory*. Cambridge, Mass.: MIT Press, 1997.

Hullot-Kentor, Robert. "Back to Adorno." *Telos*, no. 81 (Fall 1989): 5–29.

Jaggar, Alison. *Feminist Politics and Human Nature*. Totowa, N.J.: Rowman & Allanheld, 1983.

Jameson, Fredric. *Late Marxism: Adorno; or, The Persistence of the Dialectic*. London: Verso, 1990.

Jarvis, Simon. *Adorno: A Critical Introduction*. New York: Routledge, 1998.

Jay, Martin. "Is Experience Still in Crisis? Reflections on a Frankfurt School Lament." *Kriterion*, no. 100 (December 1999): 9–25.

"Mimesis and Mimetology: Adorno and Lacoue-Labarthe." In *The Semblance of Subjectivity: Essays in Adorno's Aesthetic Theory*, edited by Tom Huhn and Lambert Zuidervaart, pp. 29–53. Cambridge, Mass.: MIT Press, 1997.

Songs of Experience: Modern American and European Variations on a Universal Theme. Berkeley: University of California Press, 2005.

Kant, Immanuel. *Critique of the Power of Judgment*. Edited by Paul Guyer. Translated by Paul Guyer and Eric Matthews. Cambridge: Cambridge University Press, 2000.

Critique of Practical Reason. In *Practical Philosophy*, translated and edited by Mary J. Gregor, general introduction by Allen Wood, pp. 133–271. Cambridge: Cambridge University Press, 1996.

Keane, John. *Global Civil Society?* Cambridge: Cambridge University Press, 2003.

Keat, Russell. *Cultural Goods and the Limits of the Market*. London: Macmillan, 2000.

Klaassen, Matthew J. "The Nature of Critical Theory and Its Fate: Adorno vs. Habermas, Ltd." Master's thesis, Institute for Christian Studies, Toronto, 2005.

Knoll, Manuel. *Theodor W. Adorno: Ethik als erste Philosophie*. Munich: Wilhelm Fink, 2002.

Küng, Hans, and Karl-Josef Kuschel, eds. *A Global Ethic: The Declaration of the Parliament of the World's Religions*. New York: Continuum, 1998.

Lacy, Suzanne, ed. *Mapping the Terrain: New Genre Public Art*. Seattle: Bay Press, 1995.

Landes, Joan B. *Women and the Public Sphere in the Age of the French Revolution*. Ithaca: Cornell University Press, 1988.

Lee, Lisa Yun. *Dialectics of the Body: Corporeality in the Philosophy of T. W. Adorno*. New York: Routledge, 2005.

Lunn, Eugene. *Marxism and Modernism: An Historical Study of Lukács, Brecht, Benjamin, and Adorno*. Berkeley: University of California Press, 1982.

Macdonald, Iain. "Ethics and Authenticity: Conscience and Non-Identity in Heidegger and Adorno with a Glance at Hegel." In *Adorno and Heidegger: Philosophical Questions*, edited by Iain Macdonald and Krzysztof Ziarek. Stanford: Stanford University Press, 2007.

Margolis, Joseph. "Reconciling Analytic and Feminist Philosophy and Aesthetics." In *Feminism and Tradition in Aesthetics*, edited by Peggy Zeglin Brand and Carolyn Korsmeyer, pp. 416–30. University Park: Pennsylvania State University Press, 1995.

Martinson, Mattias. *Perseverance without Doctrine: Adorno, Self-Critique, and the Ends of Academic Theology*. Frankfurt am Main: Peter Lang, 2000.

Marx, Karl. *Capital: A Critique of Political Economy*. Vol. 1. Edited by Frederick Engels. Translated by Samuel Moore and Edward Aveling. New York: International Publishers, 1967.

Menke, Christoph. *Die Souveränität der Kunst: Ästhetische Erfahrung nach Adorno und Derrida*. Frankfurt am Main: Suhrkamp, 1991.

The Sovereignty of Art: Aesthetic Negativity in Adorno and Derrida. Translated by Neil Solomon. Cambridge, Mass.: MIT Press, 1998.

Morris, Martin. *Rethinking the Communicative Turn: Adorno, Habermas, and the Problem of Communicative Freedom*. Albany: State University of New York Press, 2001.

Müller-Doohm, Stefan. *Adorno: A Biography*. Translated by Rodney Livingstone. Cambridge: Polity Press, 2005.

Nicholsen, Shierry Weber. "*Aesthetic Theory*'s Mimesis of Walter Benjamin." In *The Semblance of Subjectivity: Essays in Adorno's Aesthetic Theory*, edited by Tom Huhn and Lambert Zuidervaart, pp. 55–91. Cambridge, Mass.: MIT Press, 1997.

O'Connor, Brian. *Adorno's Negative Dialectic: Philosophy and the Possibility of Critical Rationality*. Cambridge, Mass.: MIT Press, 2004.

Paetzold, Heinz. "Adorno's Notion of Natural Beauty: A Reconsideration." In *The Semblance of Subjectivity: Essays in Adorno's Aesthetic Theory*, edited by Tom Huhn and Lambert Zuidervaart, pp. 213–35. Cambridge, Mass.: MIT Press, 1997.

Pensky, Max. "Editor's Introduction: Adorno's Actuality." In *The Actuality of Adorno: Critical Essays on Adorno and the Postmodern*, edited by Max Pensky, pp. 1–21. Albany: State University of New York Press, 1997.

"Globalizing Theory, Theorizing Globalization: Introduction." In *Globalizing Critical Theory*, edited by Max Pensky, pp. 1–15. Lanham, Md.: Rowman & Littlefield, 2005.

Peters, Rebecca Todd. *In Search of the Good Life: The Ethics of Globalization*. New York: Continuum, 2004.

Pickford, Henry. "The Dialectic of Theory and Practice: On Late Adorno." In *Adorno: A Critical Reader*, edited by Nigel Gibson and Andrew Rubin, pp. 312–40. Oxford: Blackwell, 2002.

Schweppenhäuser, Gerhard. "Adorno's Negative Moral Philosophy." In *The Cambridge Companion to Adorno*, edited by Tom Huhn, pp. 328–53. Cambridge: Cambridge University Press, 2004.

Seerveld, Calvin. "Imaginativity." *Faith and Philosophy* 4 (January 1987): 43–58.
Sherratt, Yvonne. *Adorno's Positive Dialectic.* Cambridge: Cambridge University Press, 2002.
Taylor, Charles. *The Ethics of Authenticity.* Cambridge, Mass.: Harvard University Press, 1992.
―― *Sources of the Self: The Making of the Modern Identity.* Cambridge, Mass.: Harvard University Press, 1989.
Taylor, Ronald, ed. *Aesthetics and Politics: Debates between Bloch, Lukács, Brecht, Benjamin, Adorno.* With an afterword by Fredric Jameson. London: NLB, 1977.
Theunissen, Michael. "Negativität bei Adorno." In *Adorno-Konferenz 1983*, edited by Ludwig von Friedeburg and Jürgen Habermas, pp. 41–65. Frankfurt am Main: Suhrkamp, 1983.
Tong, Rosemarie. *Feminist Thought: A More Comprehensive Introduction.* 2d ed. Boulder, Colo.: Westview Press, 1998.
Tugendhat, Ernst. "Heidegger's Idea of Truth." Translated by Richard Wolin. In *The Heidegger Controversy: A Critical Reader*, edited by Richard Wolin, pp. 245–63. New York: Columbia University Press, 1991.
―― "Heideggers Idee von Wahrheit." In *Heidegger: Perspektiven zur Deutung seines Werks*, edited by Otto Pöggeler, pp. 286–97. Cologne: Kiepenheuer & Witsch, 1970.
―― *Der Wahrheitsbegriff bei Husserl und Heidegger.* 2d ed. Berlin: Walter de Gruyter, 1970.
Vogel, Steven. *Against Nature: The Concept of Nature in Critical Theory.* Albany: State University of New York Press, 1996.
Volf, Miroslav, and William Katerberg, eds. *The Future of Hope: Christian Tradition amid Modernity and Postmodernity.* Grand Rapids, Mich.: Eerdmans, 2004.
Vries, Hent de. *Minimal Theologies: Critiques of Secular Reason in Adorno and Levinas.* Translated by Geoffrey Hale. Baltimore: Johns Hopkins University Press, 2005.
Waugh, Joann B. "Analytic Aesthetics and Feminist Aesthetics: Neither/Nor?" In *Feminism and Tradition in Aesthetics*, edited by Peggy Zeglin Brand and Carolyn Korsmeyer, pp. 399–415. University Park: Pennsylvania State University Press, 1995.
Wellmer, Albrecht. "Communications and Emancipation: Reflections on the Linguistic Turn in Critical Theory." In *On Critical Theory*, edited by John O'Neill, pp. 231–63. New York: Seabury Press, 1976.
―― *Critical Theory of Society.* Translated by John Cumming. New York: Herder and Herder, 1971.
―― *Endgames: The Irreconcilable Nature of Modernity; Essays and Lectures.* Translated by David Midgley. Cambridge, Mass.: MIT Press, 1998.
―― *Endspiele: Die unversöhnliche Moderne; Essays und Vorträge.* Frankfurt am Main: Suhrkamp, 1993.
―― *Kritische Gesellschaftstheorie und Positivismus.* Frankfurt am Main: Suhrkamp, 1969.

The Persistence of Modernity: Essays on Aesthetics, Ethics, and Postmodernism. Translated by David Midgley. Cambridge, Mass.: MIT Press, 1991.

Whitebook, Joel. *Perversion and Utopia: A Study in Psychoanalysis and Critical Theory.* Cambridge, Mass.: MIT Press, 1995.

"The Problem of Nature in Habermas." *Telos*, no. 40 (Summer 1979): 41–69.

Wiggershaus, Rolf. *The Frankfurt School: Its History, Theories, and Political Significance.* Translated by Michael Robertson. Cambridge, Mass.: MIT Press, 1994.

Wilke, Sabine, and Heidi Schlipphacke. "Construction of a Gendered Subject: A Feminist Reading of Adorno's *Aesthetic Theory*." In *The Semblance of Subjectivity: Essays in Adorno's Aesthetic Theory*, edited by Tom Huhn and Lambert Zuidervaart, pp. 287–308. Cambridge, Mass.: MIT Press, 1997.

Young, Julian. *Heidegger's Philosophy of Art.* Cambridge: Cambridge University Press, 2001.

Zimmerman, Michael E. *Eclipse of the Self: The Development of Heidegger's Concept of Authenticity.* Rev. ed. Athens: Ohio University Press, 1986.

Zuidervaart, Lambert. *Adorno's Aesthetic Theory: The Redemption of Illusion.* Cambridge, Mass.: MIT Press, 1991.

Artistic Truth: Aesthetics, Discourse, and Imaginative Disclosure. Cambridge: Cambridge University Press, 2004.

"Autonomy, Negativity, and Illusory Transgression: Menke's Deconstruction of Adorno's Aesthetics." *Philosophy Today*, SPEP suppl. (1999): 154–68.

"Creative Border Crossing in New Public Culture." In *Literature and the Renewal of the Public Sphere*, edited by Susan VanZanten Gallagher and Mark D. Walhout, pp. 206–24. New York: St. Martin's, 2000.

"Cultural Paths and Aesthetic Signs: A Critical Hermeneutics of Aesthetic Validity." *Philosophy and Social Criticism* 29 (2003): 315–40.

"Feminist Politics and the Culture Industry: Adorno's Critique Revisited." In *Feminist Interpretations of Theodor Adorno*, edited by Renée Heberle, pp. 257–76. University Park: Pennsylvania State University Press, 2006.

"Postmodern Arts and the Birth of a Democratic Culture." In *The Arts, Community and Cultural Democracy*, edited by Lambert Zuidervaart and Henry Luttikhuizen, pp. 15–39. New York: St. Martin's, 2000.

Review of J. M. Bernstein, *Adorno: Disenchantment and Ethics. Constellations* 10 (2003): 280–3.

"Short Circuits and Market Failure: Theories of the Civic Sector." Paper presented at the Twentieth World Congress of Philosophy, Boston, 1998. http://www.bu.edu/wcp/Papers/Soci/SociZuid.htm.

"The Social Significance of Autonomous Art: Adorno and Bürger." *Journal of Aesthetics and Art Criticism* 48 (Winter 1990): 61–77.

"Theodor Adorno." In *The Stanford Encyclopedia of Philosophy*, edited by Edward N. Zalta. Summer 2003 edition. http://plato.stanford.edu/archives/sum2003/entries/adorno/.

Index

absolute, the, 54–5
Adorno, Theodor W.
  Aesthetic Theory, 16–47, 191–5
  aesthetics, 8, 16–47, 141, 191–5
  biography, 184–5
  critique of Heidegger, 85, 87, 89, 90, 95, 99
  Dialectic of Enlightenment, 12, 13, 75–6, 107–31, 137, 185–8, 190
  Jargon of Authenticity, 95
  Kierkegaard: Construction of the Aesthetic, 89
  Minima Moralia, 125, 155, 156
  Negative Dialectics, 22, 48–76, 95–101, 119, 120, 195–201
  "On the Fetish-Character in Music and the Regression of Listening," 143
  "Trying to Understand Endgame," 29
  "Why Still Philosophy?" 4–5
aesthetic dimension, 8, 24–5, 26–7, 39–42; *see also* autonomy, aesthetic; experience, aesthetic
  attitude, 23
  deferral, 20–1
  evaluation, 21
  interpretation, 21
  judgment, 24, 153
  merit, 148; *see also* imaginative cogency

negativity, 19–21, 22, 23, 25, 26–7, 29, 30–4, 38–9, 46
signs, 32–4, 40
understanding, 20, 29
aestheticism, 28
Aho, Kevin, 87
alienation, 87–91
anti-Semitism, 191
anxiety, 85
art practices, 46, 135, 140, 148
art products, 43, 148
arts organizations, 46
artworks, 17, 23, 36, 38–9, 42–4, 45–6, 70–1, 133, 135, 138, 142, 148, 191–5
authentication, 14, 67–9, 87, 91, 95, 98–9, 99–101, 101–6, 168; *see also* authenticity; justification
authenticity
  artistic, 43–4
  existential, 68, 79, 80–95, 98–9, 102, 166
autonomy, 29, 153, 158, 160–1
  aesthetic, 23, 25, 26–7, 30–4, 37, 38–47
  artistic, 10, 13–15, 17–18, 19, 23, 25, 26, 34–7, 38–47, 132
  internal, 13–15, 17, 137, 138, 148, 153
  personal, 12, 137, 153
  societal, 137, 138, 148, 150, 152, 153

avant-garde, 133

Baca, Judith, 149
Bacon, Francis, 186
Bauman, Zygmunt, 178
beauty, 25, 38, 137
Becker-Schmidt, Regina, 136
Beckett, Samuel, 29, 194
Beethoven, Ludwig von, 65
Bell, Clive, 134
Benhabib, Seyla, 110, 133, 157
Benjamin, Walter, 35, 36, 133, 192, 194
Berg, Alban, 183
Berman, Russell, 157, 158, 161
Bernstein, J. M., 8–9, 72, 74, 96, 114, 117, 157, 170
Brand, Peggy Zeglin, 148
Brantlinger, Patrick, 145
Brecht, Bertolt, 133, 194
Brunkhorst, Hauke, 167–8, 169
Bürger, Peter, 17
Butler, Judith, 133

capitalism, 115, 116–17, 120, 125–6, 129–30, 135, 142, 145, 153, 163–5, 168–70, 188–91, 194
Carman, Taylor, 92–4
Carroll, Noël, 32, 144
categorical imperative, 60–1, 69, 179–80
Cavell, Stanley, 159
Chicago, Judy, 149
Christianity, 56–7; *see also* theology
civil society, 127, 128, 150, 165, 166, 167–8, 169, 171–5
collectivity, 161–2, 175–7
colonization of the lifeworld, 120, 128, 147
commodity, 137, 142, 145
  fetishism, 188–9
communication, 66–9, 100–1; *see also* intersubjectivity
concept, 96, 114–15, 116, 117, 195
conscience, 82, 85, 87–90, 94
constellation, 65, 197
control, 121–2, 125, 127
Cook, Deborah, 9, 16, 68–9, 71, 120, 139, 144
Cornelius, Hans, 184

Cornell, Drucilla, 133, 157
countereconomy, 148
counterpolitics, 148
creative interpretation, 39–40
cultural politics, 133, 145
culture, 55, 161
culture industry, 9–10, 13–15, 16–17, 132–54

Dahlstrom, Daniel O., 80, 81
Danto, Arthur, 32, 46
death, 22–3, 60, 62–3, 81
democracy, 166–8, 168–70
Derrida, Jacques, 19, 21–2, 28, 33, 34, 178
determinate negation, 115–16, 117, 187–8
Devereaux, Mary, 134–5, 146
Dewey, John, 166
dialectics, 115, 198–9
differential transformation, *see* transformation
differentiation of society, 121, 127, 128, 130, 146, 168–9, 174; *see also* transformation
disclosedness (*Erschlossenheit*), 78–95, 101
disclosure, 102–6, 168–70, 170; *see also* truth
discourse, 23, 34, 37
discoveredness (*Entdecktheit*), 78, 79
discursive practices, 103–5
domination, 9, 12–13, 114, 116–17, 120–4, 125–31, 186–7
Dreyfus, Hubert L., 92
Dufrenne, Mikel, 24

economics, 120, 125–6, 127, 128, 145–52, 161, 163–5, 168–70, 171–5; *see also* capitalism
  third sector, 172–5
elitism, 96–8, 100, 158, 159, 165
end of art, 46
enlightenment, 111–12, 114–15, 115–18, 188–9
ennoetism, 52
epistemology, 53–4, 61
ethics, 15, 60–1, 62–4, 89–91, 155–81, 199–201; *see also* morality
  discourse, 157

Index

emphatic, 157–62
postmodern, 178
of resistance, 159, 180
social, 177–81
evil, societal, 58–9, 61–2, 74–6, 118, 120, 165
exchange principle, 54, 73, 115, 190, 196
experience
  aesthetic, 18–19, 19–21, 21–3, 24, 25, 26–7, 28, 29, 32–4, 38–9, 40
  emphatic, 11–12, 22, 95–101, 103, 166, 168
  metaphysical, 52, 53, 58, 59–65, 96; *see also* metaphysics
  philosophical, 66–9, 74–5, 95, 96–8, 99–101, 195
exploitation, 121, 124, 125–6, 127
exploration, 39–40

Felski, Rita, 152
feminism, 10, 13–14, 133–8, 145
  radical, 136, 147
  socialist, 136
Feuerbach, Ludwig, 199
fidelity, 102, 104, 105
finitude, 64
formalism, 134
form-content dialectic, 138, 194
Frankfurt School, 4–7, 184
Fraser, Nancy, 150

Gadamer, Hans-Georg, 183
genius, 24
German idealism, 49–50, 195
Geuss, Raymond, 67–8, 155
global ethic, 170, 171
globalization, 9, 125–31, 162–5, 167–8, 168–70, 171–5
Goudzwaard, Bob, 129–30
Gracyk, Theodore, 144
Greenberg, Clement, 134
Guignon, Charles, 92
guilt, 87–90
Gutiérrez, Gustavo, 57

Habermas, Jürgen, 4–6, 8, 9, 12–13, 28, 46, 50, 64, 67, 74, 105, 108–9, 110–13, 114, 118–20, 121, 125–31, 147, 151, 152, 157, 183
Hammer, Espen, 9–10, 59, 142, 157, 158–61, 162–5, 170
Hansen, Miriam, 149
happiness, 65
Hare, John, 180
Haskins, Casey, 24
Hegel, G. W. F., 25, 38, 50, 54–5, 60, 111, 116–17, 120, 125, 161, 163, 167, 176, 187, 192, 195–8
Heidegger, Martin, 12, 50, 63, 78–95, 98–9, 101, 102–3, 104, 105–6, 111, 141, 166, 183, 195
Held, David, 163, 167
Honneth, Axel, 75–6, 109–10, 157, 161
hope, 11, 12–13, 56–8, 59, 61–5, 70–6, 164–5
Horkheimer, Max, 107–31, 143, 183, 184–7
hypercommercialization, 145, 147, 191

idealism, 111
identity thinking, 54–5, 61
imagination, 33, 39–42
imaginative cogency, 41, 148–50
imaginative disclosure, 43–4
immortality, 62–3
imperialism, 128
import, artistic, 42, 191, 193–5
Institute of Social Research, 184
institutions, 104, 127
  cultural, 102–3, 127, 146, 150, 153
  social, 102–3, 125
integrity, 42–4
internalization, 161
intersubjectivity, 66–9, 93–4, 97–8, 122; *see also* communication; public world

Jameson, Fredric, 156, 159, 160
jargon of ethicality, 157; *see also* ethics
Jarvis, Simon, 8, 53, 71, 195
Jay, Martin, 96
jazz, 144
Judaism, 56–7, 116; *see also* theology

justice, 128, 129; *see also* societal principles
justification, 67–9, 100–1, 103–5; *see also* authentication

Kant, Immanuel, 24, 25, 31, 32, 40, 49, 51, 52, 54, 60, 62–4, 65, 137, 144, 151, 153, 176, 179, 180, 192, 195–8
Keane, John, 169, 171–2, 174
Keat, Russell, 146
Klaassen, Matthew, 119
Knoll, Manuel, 175–6
Kristeva, Julia, 34
Küng, Hans, 169, 172

Landes, Joan, 151
Lazarsfeld, Paul, 191
Leppert, Richard, 143
Lukács, Georg, 27, 111, 133, 160, 189–90, 194

Macdonald, Iain, 89–91
Mallarmé, Stéphane, 24
Margolis, Joseph, 154
Marx, Karl, 27, 51, 75, 89, 120, 125–6, 137, 143, 144, 151, 162, 163, 164, 167, 188–9, 192, 199–200
Marxism, 159, 160
mass culture, *see* culture industry
McGrew, Anthony, 163, 167
mediation, 91–4, 100
Menke, Christoph, 8, 10–11, 18–38
metacritique, 49, 51–4, 195
metaphysics, 8, 11–12, 48–76, 199–201
micrology, 55, 65, 66
mode of production, 35
modern art, 27, 28–9, 191–5
modernity, 185–8
modernization, 128; *see also* rationalization
Moltmann, Jürgen, 57
morality, 62; *see also* ethics
Morris, Martin, 9, 74, 170
myth, 115–18, 187–8

need, 55–6, 64, 74–5, 123
new genre public art, 45

Nietzsche, Friedrich, 59, 111–12
nonconceptual, the, 96, 196; *see also* nonidentical
nonidentical, the, 34, 61, 63, 66, 70, 71–2, 75, 90, 114, 162–3, 196–7
normativity, 129–30, 161–2, 175–7; *see also* societal principles; validity
aesthetic, 41, 43
social, 94

object, 29–30, 75, 96, 119–20, 162–3
priority of, 40, 54–8, 63, 71–2, 197–8
O'Connor, Brian, 95, 114
Okrent, Mark, 81

patriarchy, 136
Pensky, Max, 2, 126
performance fetishism, 147
Peters, Rebecca Todd, 169–70, 172, 174
philosophy, 4–7, 200
political theory, 160–2, 165; *see also* politics
politics, 15, 127, 128, 146, 155–81; *see also* cultural politics
Pollock, Friedrich, 160, 190
Popper, Karl, 183
postaesthetic subversion, 21–3
presentation (*Darstellung*), 39–40, 197
progress, 186
public sphere, 46, 150, 160–2, 171
public world, 86–7, 87–9, 93–4, 97–8, 100–1; *see also* intersubjectivity
purposiveness without purpose, 137, 142, 153

rationality, 27, 28–9, 41–2, 96, 111–12, 117
aesthetic, 41–2
discursive, 23, 35
instrumental, 111–17, 186
rationalization, 28, 110, 189; *see also* modernization
reason, *see* rationality
reification, 109, 111, 118–19, 189–90

Index

releasement (*Gelassenheit*), 87
remembrance of nature, 12, 111, 112–24
repression, 121, 122–3, 125, 127
resoluteness (*Entschlossenheit*), 82–4, 84–7, 91, 101
resourcefulness, 128, 129; *see also* societal principles
responsibility, 90
rock music, 144
Rorty, Richard, 81

sacrifice, 122
Schelling, F. W. J., 28
Scholem, Gershom, 56–7
Schweppenhäuser, Gerhard, 178
scientism, 115–16
Seerveld, Calvin, 38
self-reflection, 188
self-relation, 84–7
semblance (*Schein*), 51
Sherratt, Yvonne, 71
Short, Jon, 69–70
significance, 43–4
societal principles, 74–6, 102, 118, 120, 125–31, 169, 180
solidarity, 128, 129; *see also* societal principles
speech acts, 32–4
spirit (*Geist*), 55, 63–4
structural integration, 169–70
subject, 54, 73, 96, 119–20
sublimation, 123–4
subsystems, 128; *see also* differentiation of society
suffering, 53, 58, 59–65, 66, 67–8, 68–9, 69–70, 74–6, 197
symbol, 115–16

Taylor, Charles, 176
technique, 29, 35, 36–8, 144
technology, 35, 36–7

temporality, 81
theology, 56–8
Theunissen, Michael, 51–2
Tillich, Paul, 56–7, 183
Tocqueville, Alexis de, 140, 150
transformation
  differential, 12, 14, 125–31, 169–70
  social, 54, 72–4, 118
transgression, 30–8, 162
truth, 12, 51–4, 61–2, 66–9, 70, 77–106, 111, 168, 200; *see also* disclosure
  artistic, 42–4, 194–5
  propositional, 78–95
truth content (*Wahrheitsgehalt*), 29, 42, 44, 49, 53, 194–5
Tugendhat, Ernst, 78–80, 80–1

universities, 173–4
utopia, 74–6

validity, 104, 105
  aesthetic, 41, 43
validity claims, 103, 104
verification, 103–5
violence, 121–4
Vogel, Steven, 119
Vollenhoven, Dirk, 52
voluntarism, 87
voluntary associations, 140, 150

Weber, Max, 110, 189
Wellmer, Albrecht, 11–12, 49–58, 64, 70, 74, 109–10, 157
Whitebook, Joel, 119, 123–4
Wilke, Sabine, 17

Young, Julian, 87

Zimmerman, Michael, 87

For EU product safety concerns, contact us at Calle de José Abascal, 56–1º,
28003 Madrid, Spain or eugpsr@cambridge.org.

www.ingramcontent.com/pod-product-compliance
Ingram Content Group UK Ltd.
Pitfield, Milton Keynes, MK11 3LW, UK
UKHW020328140625
459647UK00018B/2066